Muscles, Nerves, and Pain

Wilfred A. Nix

Muscles, Nerves, and Pain

A Guide to Diagnosis, Pain Concepts, and Therapy

Second edition

With 102 figures

 Springer

Nix, Wilfred A.
Professor of Neurology
Department of Neurology
University Medical Center
Johannes Gutenberg University
Mainz, Germany

ISBN 978-3-662-53718-3   978-3-662-53719-0 (eBook)
DOI 10.1007/978-3-662-53719-0

Springer

Cover Design: deblik Berlin
Graphic design and layout: Berthold Matthäus, Mühltal
Cover illustration: © Prof. Nix, Mainz
Illustrations: Fotosatz-Service Köhler GmbH – Reinhold Schöberl, Würzburg

Printed on acid-free paper

This Springer imprint is published by Springer Nature
The registered company is Springer-Verlag GmbH Germany
The registered company address is: Heidelberger Platz 3, 14197 Berlin, Germany

Acknowledgments

I would like to thank all those who encouraged me to write this book and who helped in editing and getting it published. For the first edition I would particularly like to thank Dr. Dieter Bonke, who coordinated much of the work that had to be done, and Berthold Matthäus for the graphic design and layout. For the second edition I thank the continuing support of Springer represented by Diana Kraplow, Astrid Horlacher, and Isabella Athanassiou. Their help guided me through the book's revision.

I am also indebted to my patients, who were a constand motivation while writing the book. Finally I would like to thank my wife, Brigitte, for her patience and constant support and care.

Contents

Introduction

Diagnosing nerve and muscle lesions and the pain associated with them is generally considered difficult, as they display a wide variety of symptoms and the mere thought of the extensive anatomical details is a deterrent. However, these reservations are unfounded. A targeted patient history and practical examination techniques that can be learned make it possible to recognize many lesions. This is also necessary because disorders such as back pain, sensory disorders, and muscle weakness are not rare, and since many patients suffer from them, they demand a diagnostic response.

Nerves and muscles are at the forefront of diagnostic interest in peripheral neurological disorders. This is understandable, as muscle paralysis is always a dramatic symptom and nerves are exposed to a great deal of potential damage along the long course from the spinal cord to muscle or skin. This yields a great variety of lesion sites that must be considered in daily practice. For example, we have to decide whether muscle paralysis is caused by disc damage or only by a nerve lesion. Very frequently, for example, pain brings the patient to the doctor, who then must differentiate between myogelosis in the arm or entrapment syndrome in the carpal tunnel. It is necessary to distinguish between acute and chronic pain, as different peripheral or central nerve mechanisms are involved and they determine the required therapy options. Pain can be nociceptive or neuropathic. Occasionally both components play a role – a "mixed pain" concept that is described here.

Depending on the suspected diagnosis, different procedures are required – and at different speeds. For example, immediate medication is needed for trigeminal neuralgia before any further examination. By contrast, the suspicion of an acute herniated disc requires targeted, technical diagnostics before any therapy. Diagnosing peroneal nerve paresis, on the other hand, is not as urgent. But since it can be mistaken for a herniated disc, careful differentiation must first be made between these two lesions in the peripheral area. Back or neck pain in the form of lumbago or a cervical syndrome is a very frequent acute or chronic symptom in which the mixed pain concept can be helpful in understanding the illness as well as treating it. Facial nerve paralysis can also make a differential diagnosis difficult when you need to decide between a peripheral or central facial nerve paresis.

In every case, the first physician to see the patient sets the course with his suspected diagnosis by planning and arranging further tests. In making his diagnosis, he is aided by the fact that damage to peripheral nerves always occurs systematically depending on the location of the lesion. The loss of motor, sensory, and vegetative function is different for root damage than for a lesion in the plexus or further toward the periphery. This can be helpful in solving the problem cycle caused by the damage.

More difficult are painful disorders in the functional relation between joints and their associated muscles, which can imitate nerve irritations. This book will also offer assistance in these situations by presenting the newest ideas on how symptoms arise and the ensuing knowledge of diagnostics and therapy. Furthermore, familiar know-how is refreshed in pictures and practice-related texts. This book is not intended as a textbook, but rather as a guideline for routine practice, and its numerous illustrations can be helpful for explaining the situation to the patient.

Wilfred A. Nix

Basics of anatomy

1

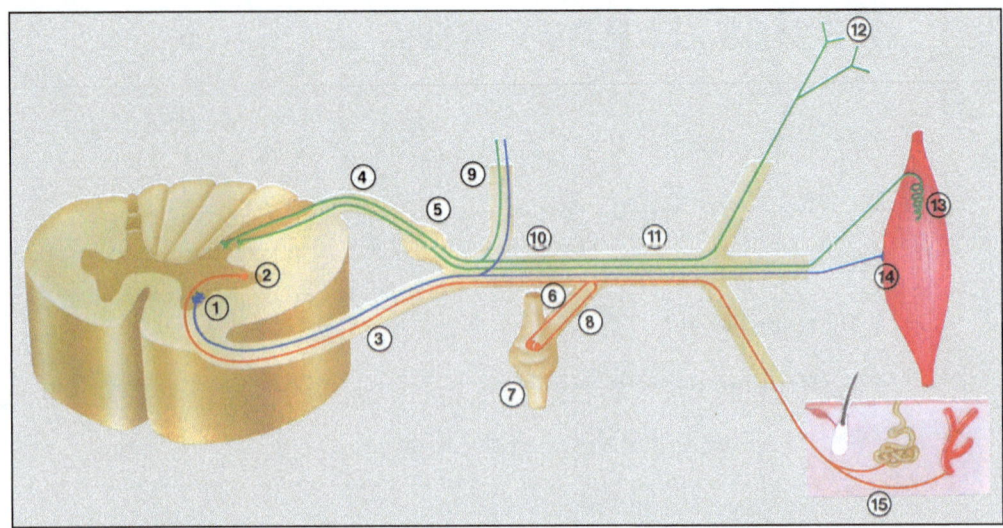

Figure 1 Peripheral nerve. The peripheral nerve conducts impulses from the spinal cord to its target organs in the periphery, skin, and muscles. From there, receptors send information back to the spinal cord. This exchange of information takes place along the following pathways: blue = motor, red = vegetative, green = sensory. 1) Ventral horn cell, 2) Sympathetic lateral horn, 3) Ventral root, 4) Dorsal root, 5) Spinal ganglion, 6) Ramus communicans griseus, 7) Sympathicus trunk, 8) Ramus communicans albus, 9) Dorsal spinal nerve, 10) Ventral spinal nerve, 11) Plexus, peripheral nerve, 12) Receptors for pressure, vibration, temperature, pain, 13) Muscle spindle, 14) Neuromuscular synapse, 15) Vegetative innervation of hair, sweat glands, and vessel

Peripheral nerves contain motor, sensory, and vegetative fibers that are strictly separate, but in close proximity to each other (Figure 1).

As the peripheral nerve system, they connect the central system with its target organs and thus make reaction possible. They contribute to the success of the complex organization of body functions by being limited strictly to their task of transmitting information and do not generate any information themselves. If one of these working conditions is not adhered to, disorders develop. Such misdirected impulses in the motor system can lead to a facial nerve spasm, in the sensory system to trigeminal neuralgia, and a combination can also cause the development of pain syndromes.

1.1   Motor system

Voluntary motor activity always proceeds centrifugally from the motor cortex cell. From there, the action impulse proceeds in the central nervous system via the axon cylinder of the cell toward the periphery, that is, to brain cell nuclei or the motor ventral horn cell of the spinal cord. This distance is called the 1st motor neuron. The 2nd motor neuron is the ventral horn cell. It releases its axon cylinder via the ventral spinal root to the plexus. It splits to the peripheral nerve, which, after a long distance, ends at the neuromuscular synapse of the striated muscle. Because the 2nd neuron arises in the central nerve system, damage occurring there, such as an insult, can affect the peripheral nerve. Thus, in a brain stem infarction, peripheral facial nerve paresis can arise as well as muscle paralysis from damage to the ventral horn cell in the spinal cord (Figure 2).

1.2   Sensory system

The peripheral sensory neuron is fundamentally different from the motor neuron. Its cell body is located outside of the spinal cord in the spinal ganglion and has two neurites. The sensory pseudo-unipolar neuron cell bodies are located in the dorsal root ganglion, and their afferent and efferent projections come

Figure 2 Motor nerve cell and signs of degeneration.
1 Cell body with peripheral nucleus and organelles of the cytoplasm, 2 Axon with axon cylinder and myelin sheath, 3 Degenerating axon with disintegrated Schwann cells, 4, 5 Dendrite with microglial cell, 6 Astroglial cell, 7 Fibrocytes and collagen fibers, 8 Capillaries, 9 Tissue-clearing phagocyte

off the dorsal roots housed in the intraforaminal space. Sensory information generated by peripheral receptors initially proceeds centripetally via the axon cylinder in the plexus to the ventral root. From the ganglion where the cell body of the nerve cell is located, an axon cylinder proceeds via the central portion of the dorsal root into the spinal cord. There the nerve is linked by synapses with segmental as well as ascending and descending tracts. The large amount of incoming information is initially processed at the spinal level, so that a large portion of the stimuli going to the brain are already analyzed there. Knowledge of these complex mechanisms facilitates understanding of the significance of peripheral, spinal, and central factors in the pain phenomenon.

Afferent sensory neurons are adequately stimulated only by certain stimuli according to the theory of specificity. The following applies to the individual groups of fibers:

- The initial, pricking, piercing components of pain sensation are conducted via A-δ-afferents.
- The secondary duller, burning sensation of the pain component occurs via nociceptive C-afferents.
- Cold receptors on A-δ-afferents are responsible for the sensation of cold.
- Pressure sensation is also detected via slowly adapting receptors on A-β-afferents.
- A-β-afferents have vibration receptors for sensing vibrations.

Sensations and thus also pain sensation result when a stimulus does not selectively activate only one receptor. Instead, various receptors are activated simultaneously to different extents. This afferent information then proceeds via another population of neurons in the spinal cord and brain stem (second neuron) to the thalamus (third neuron) and on to various cortex areas of the cerebral cortex (fourth neuron). It should be noted that there is no clear allocation, but that interneurons and projection neurons are also involved in conduction. The central nociceptive neurons are classified according to their response characteristics and stimulus threshold to peripheral stimuli. A differentiation must be made between low-threshold neurons and high-threshold neurons and those which have a low threshold but a larger receptive field and therefore are called wide dynamic range neurons (WDR). The important phenomenon of wind-up (see Sect. 6.2.4) is generated in the WDR neurons in response to C fiber stimulation and is one form of central sensitization. But wind-up can also be generated by A-β fiber-induced pain called allodynia. This is pain due to a stimulus that does not normally provoke pain. An example for this is pain in postherpetic neuralgia, when lightly touching hypoalgesic skin evokes burning pain. In this situation, morphological reorganization of synaptic contacts has taken place in the spinal cord. A-β afferents are now connected by heterosynaptic contact to second-order nociceptive neurons.

In the clinical setting, quantitative sensory testing (QST) can be used to examine and measure the detection threshold of accurately calibrated sensory stimuli. Vibratory, thermal, or painful stimuli are

1

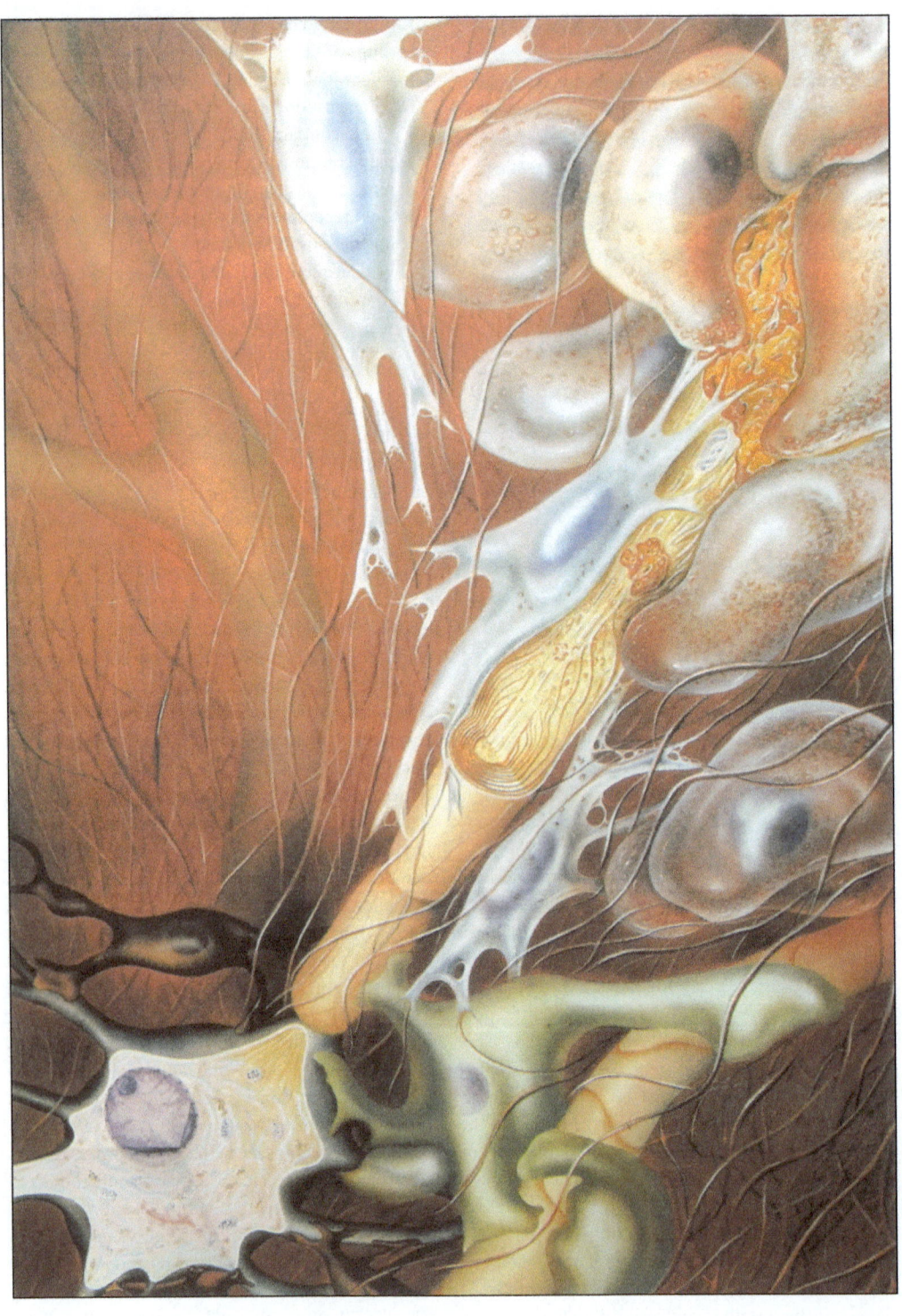

often chosen because they relate to distinct neuro-anatomic pathways with discrete fiber populations. If QST reveals hyperalgesia only in connection with mechanical stimuli, this can be taken as an indicator for central sensitization (see Sect. 7, "Pain mechanics").

1.3   Vegetative system

The most significant vegetative function in peripheral nerve damage, sweating, is regulated by the sympathicus. Nerve tracts pass through the hypothalamus from cortical centers and move in a lateral tract in caudal direction without crossing over. Cell bodies located in the lateral horn chain give axons to the sympathetic trunk via the ventral root and the rami communicantes albi. There they switch to postganglionic nerve cells whose axons lead via the ramus communicans griseus in close proximity to the sensory nerve fibers of the skin to innervate the sweat glands there. Innervation of the piloerector muscle enables the goose bump reaction. In other emotionally charged situations, skin color and temperature can be changed by the regulation of blood vessel diameters. It has recently become known that vegetative fibers have secretor functions through which they can influence the trophic supply of the skin and also that they play an important role in pain.

1.4   Neuron

Nerve cells have their own cell body from which processes arise. These frequently extensive and highly branched cell processes are incapable of surviving without contact to the cell body. The motor nerve cell has several short processes called dendrites, which allow communication with other nerve cells in the spinal cord. The cell can form new processes at any time if necessary. The axon is the longest process; it leaves the spinal cord to make contact with a muscle. An axon consists of an axon cylinder, a fluid-filled neurilemma in which nutrients, enzymes, chemical signals, and structural proteins flow distally from the cell body. There is also a reverse axoplasmic flow that transports information from the periphery to the cell body (Figure 3a).

The following example explains the regenerative capacity of the nerve cell: if the nerve were about 1.80 m in diameter, the axon would be 2.5–5 cm thick. The distance from the lumbar spinal cord to the small foot muscles that the axon bridges would be equivalent to 5 km. The flow velocity varies in both directions between a few and 1 mm per day. This velocity takes on importance when a nerve resprouts after an injury, as the transport speed at which structural proteins arrive determines the repair time of the nerve. Changes in the axoplasmic flow also play an important role in the development of polyneuropathies and are described in more detail in Sect. 8.2.2, "Axonopathies" (Figure 3b).

In addition to material transport that serves to maintain the nerve structure itself, its task is in particular to transmit electric signals.

This information transfer is necessary for the functioning of the organism as a whole. When signals are transmitted, distances of up to 1 m must occasionally be bridged. To be able to conduct electric impulses quickly over these long distances, the axon is coated with a special sheath, the myelin. The thickness of the myelin coating distinguishes medullated from non-medullated nerve fibers. The thick medullated motor fibers conduct fastest (60–120 m/s), the non-medullated pain and sympathetic fibers only slowly (2–0.3 m/s).

Every peripheral nerve axon is sheathed from the root zone to the distal axonal termination site by a continuous basal lamina of sequentially placed Schwann cells. A myelin sheath is expressed by Schwann cells on the larger motor fibers and the mid-sized sensory fibers. The smallest polymodal nociceptive fibers are unmyelinated. The lipid-rich myelin material enables saltatory conduction of action potentials between nodes of Ranvier – the unmyelinated axon between adjacent myelinated Schwann cells. The width of this internode distance plays an important role for the velocity of impulse conduction along the nerve. As the potentials are conducted by saltatory conduction, that is leaping over the myelin sheath of what is called an internode, they reach the destination more quickly than by continually traveling along the axon as on an unmyelinated axon. Each node of Ranvier has an increased density of sodium channels compared with the paranodal area, which has a higher concen-

1

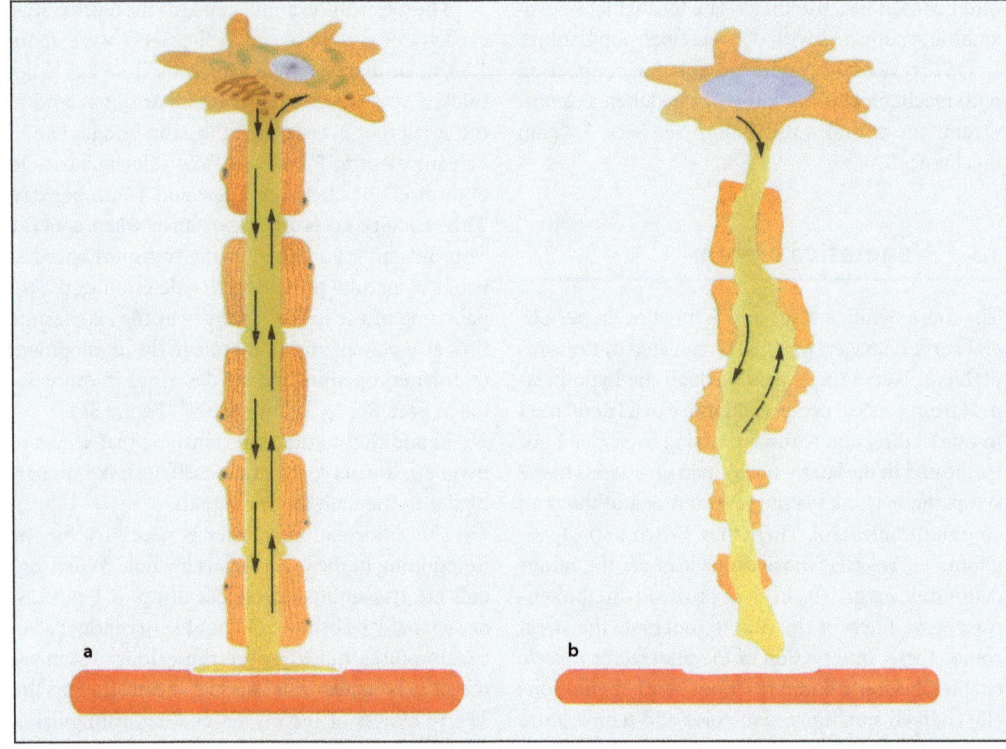

Figure 3 Axoplasmic flow. In the axon cylinder of the nerve, chemical substances are transported in a fluid (axoplasmic flow) from the cell body to the nerve ending and from there back to the cell body. The flow supplies the axon and the muscle with nutrients (**a**). Problems in the axoplasmic flow damage the axon cylinder, which then dies back from distal to proximal end; the denervated muscle atrophies (**b**). This situation is often found with polyneuropathies and is known as "dying back neuropathy"

tration of potassium channels. The nodes of Ranvier act as a relay station for saltatory impulse conduction. The basal lamina is preserved when a Schwann cell degenerates and a duplicate basal lamina is formed by new Schwann cells. The basal lamina has a crucial function in axonal regeneration because it degenerates very slowly after Schwann cell death. Therefore, it serves as a conduit for nerve fiber regeneration in lesions where the axon has not been transected (Hanke-Büngner bands). After a nerve lesion and axonal regeneration, demyelination, and remyelination, the myelin sheath can regenerate but the sheath will be thinner and the internode distances will be much smaller (Figure 7). These changes explain the slower conduction velocity of these nerves. However, the slowdown does not necessarily mean that clinical function must be poorer.

1.5   Peripheral nerve

In the peripheral nerve are a number of axons that are bundled together by connective tissue septa, the perineurium. Within the fascicles the individual axons are enveloped by the endoneurium; the entire nerve is enveloped by the epineurium. The arrangement and number of fascicles determined by the septa do not remain the same throughout the length of the nerve (Figure 4). These anatomical conditions alone can give rise to considerable problems when suturing the fascicle of a nerve, especially if part of the nerve is missing or must be resected. For example, a proximal nerve ending with only two fascicles must then be anastomized with five fascicles in the distal stump. Then unrelated nerve tracts are usually connected and this may cause unwanted

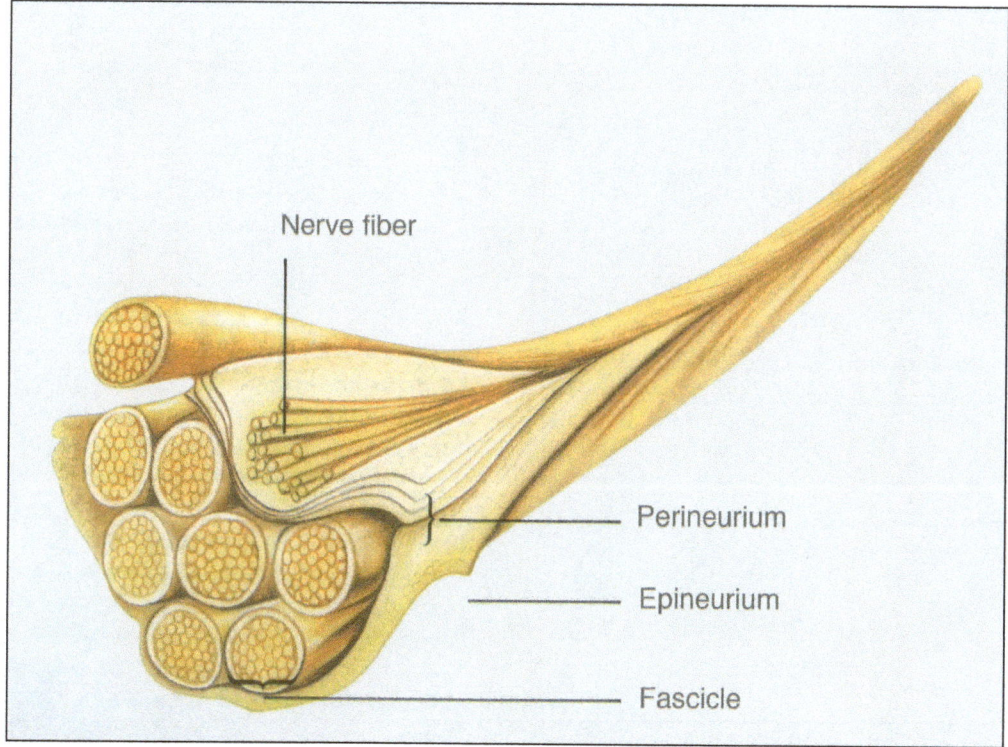

Figure 4 Nerve structure. The peripheral nerve consists of individual fascicles that sheathe the epineurium (see also Figure 83)

aberrant reinnervation with motor and sensory function damage.

Function disorders in nerves are frequently associated with morphological changes. External damage such as sharp or blunt trauma, ischemia, or inflammations can change either the insulating myelin alone (Table 1), the width of the nodes of Ranvier, or they can sever the axons and connective tissue coating. In the case of polyneuropathies, the damage is from metabolic (diabetic) and toxic (chemotherapy) noxa. They have a primary influence on metabolism in the nerve cell or obstruct the axoplasmic flow. As a result of the lack of nutrition and substrate substances, the axon degenerates secondarily. As the axon segments located furthest distal from the cell body are most poorly supplied, they are the first to be damaged. This is also the reason why polyneuropathies usually occur first in distal areas of the body. Immunological damage can cause the Na, K, and Cl channels important for conducting electrical impulses to be blocked. The fact that

this is reversible explains the prompt effect of clinical improvement following plasmapheresis for Guillain-Barré syndrome. Removing the blockage of the electrolyte channels caused by antibodies restores the conductivity of the nerve and thus its function. The level of severity of the damage also determines the potential for regeneration. Neurapraxia first causes a loss of conductivity in the nerve;

Table 1 Classification of nerve injuries. The various traumatic lesions are designated as follows if certain (+) anatomical structures are destroyed:

	Neura-praxia	Axonot-mesis	Neurot-mesis
Myelin	+	+	+
Axon		+	+
Endo-, peri-, or epineurium			+

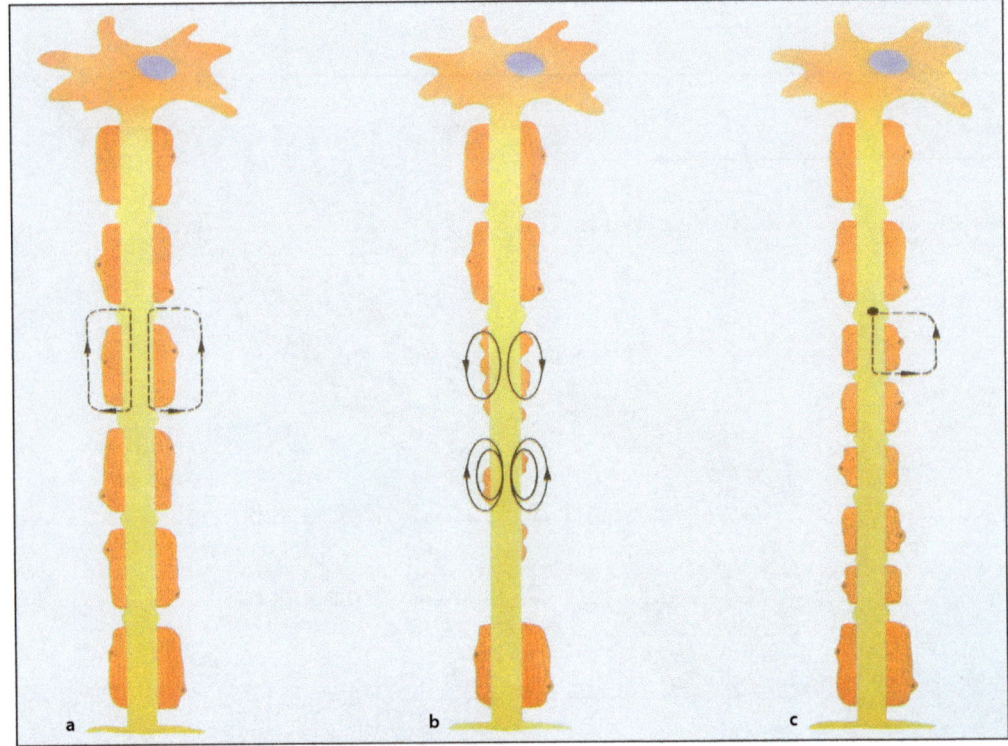

Figure 5 Nerve conduction in the myelinated nerve. The axon is responsible for conducting electric signals. A considerable acceleration of conduction velocity is possible through saltatory conduction (**a**). Degenerating myelin sheaths or the naked axon can conduct electric stimulation only through slow continuous conduction (**b**). The shorter and thinner internodes after remyelination slow down the originally faster conduction capacity of the nerve (**c**)

this lesion has the best prognosis of all types of damage. If the axon cylinder is sound, the demyelinated axon segment regains conductivity – although reduced – after remyelination. By contrast, axonotmesis means severe damage to the axon process and causes Wallerian degeneration of the nerve cell. The distal axon segment disconnected from the nerve cell dies. It uses up the nutrients stored in its axoplasm and the myelin dissolves and degenerates. The axon looses conductivity within 4–10 days. The nerve can then regain function only if it resprouts over the entire degenerated segment. As long as trophic substances are still available in the distal nerve segment, the membrane potential of the adjacent muscle fiber is not changed. When the substances are lacking, the muscle membrane potential becomes unstable, resulting in spontaneous depolarization. This is used for diagnostic purposes in

electromyography (EMG); the muscle, which now also shows trophic denervation, exhibits spontaneous activity.

1.6   Regeneration of the peripheral nerves

The interruption of continuity of the axon cylinder leads to massive changes in the entire nerve cell. Following the lesion, other nerve cells withdraw their dendrites from the surface of the damaged cell and isolate it from the synaptic cluster in the central nervous system (CNS). In the cell body, the Nissl bodies degenerate (tigrolysis), the body itself swells and its nucleus becomes marginal. These changes are signs of increased metabolic activity used by the cell to organize its survival and use all metabolic

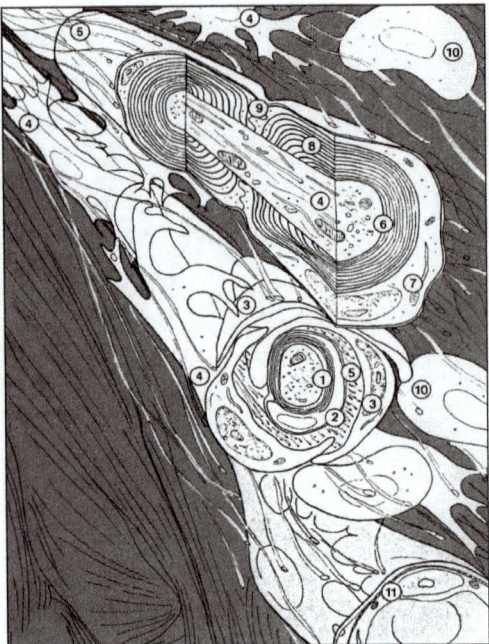

Figure 6 Normal and regenerating nerve. 1 Regenerated nerve with a thin myelin sheath, 2 Neurilemma, basal membrane of a Schwann cell, 3 Superfluous Schwann cell, 4 Axoplasm of the axon cylinder, 5 Endoneural interstitium with collagen fiber, 6 Normal nerve, 7 Schwann cell, 8 Myelin sheath coating, 9 Node of Ranvier, bordering an internode, 10 Monocyte, 11 Capillary with erythrocyte

means for its own regeneration. Peripheral to the proximal axon (Figure 7) axoplasm is exuded from the stump, which first forms a growth bud, from which axon sprouts branch out to grow toward the periphery. The severed axon breaks down and it and the myelin are cleared away by macrophages. From the remains of the axon, histiocytes, and proliferating Schwann cells, Hanke-Büngner bands are soon formed, into which the axon sprouts grow and which are the indispensable conduction pathways to the sensory target organ or muscle during the regeneration and reinnervation process (Figure 6). The growth rate is different in one extremity from proximal to distal end, and is about 5–1 mm per day. If access to the pathways in the periphery is blocked by scar tissue, the growing nerve knots up and a neuroma develops. Important for diagnostics is that the growing tips of the axon sprouts are sensitive to pressure. Light mechanical irritation can elicit

action potentials on the electrically unstable nerve fiber membranes, known as Tinel's sign. Slight percussion along the course of the nerve eliciting a Tinel sign allows the progress of reinnervation or formation of a neuroma to be followed using simple clinical means.

1.7   Denervation and reinnervation

Muscle paralysis can arise from various causes. For example, demyelinating lesions prevent voluntary muscle activation by blocking conduction. The result is atrophy due to inactivity. Since the muscle still has an intact axon and the axoplasmic flow is supplied by trophic factors, it is not subject to denervation atrophy and shows no spontaneous activity in EMG. This situation changes dramatically if the axon of the nerve is also severed. This then quickly leads to denervation atrophy and the muscle develops the EMG signs of spontaneous activity. To restore the nerve-muscle bundle, the sprouting nerve must bridge regeneration distances of varying lengths depending on where the lesion is. The further distal the lesion and thus the shorter the distance, the more favorable the reinnervation result usually is.

In radicular lesions or in polyneuropathies, only individual nerve fibers within the entire nerve degenerate, with the result that denervated muscle fibers are found next to innervated ones in the muscle. The intact axons immediately begin to innervate these denervated muscle fibers. To do so, the nerve sprouts small nerve fibers either from its terminal section (terminal sprouting) in the end plate region or in the area of the nodes of Ranvier (collateral sprouting), which quickly form an end plate at the denervated muscle and thus control it. The individual nerve fiber forms a motor unit with the muscle fiber attached to it, which can expand considerably in size during such reinnervation processes. The activation of the motor unit is accompanied by electric discharge of the muscle fibers that can be recorded with an EMG needle in an electromyogram. The EMG is thus very useful in assessing changes in a motor unit.

The main problem of nerves severed far from the muscle is that in order to regenerate, they must

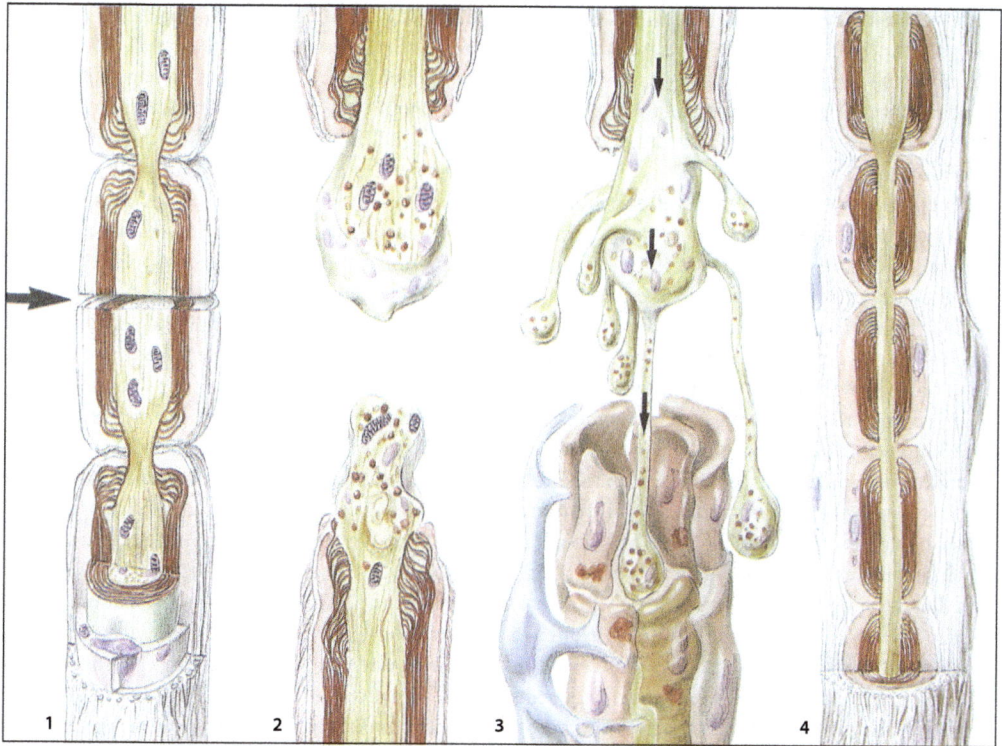

Figure 7 Sequelae of nerve injuries. 1 Sharp severance of an axon, 2 Axoplasm exudes from the proximal stump and forms the growth bud, the distal stump degenerates, the Schwann cells form a conduction pathway, 3 Axon sprouts from the growth bud grow either into a conduction pathway or miss a conduction structure, 4 The regenerating nerve has smaller and short er internodes. This affects nerve conduction velocity

find the right conduction pathway to the periphery. A specific nerve cable must find precisely the muscle fiber that was innervated by it before being damaged. Usually this exact readaptation is not successful, as sequelae after severe facial nerve pareses show. Aberrant regeneration can result in synkinesis. This is expressed, for example, in that nerve fibers that previously supplied the orbicularis oculi muscle now grow to the orbicularis oris muscle and innervate it. Every time the patient closes his eyelid, this movement is linked with a synchronous muscle contraction at the corner of the mouth. Synkinesis such as this is especially problematic after nerve damage to the arm. Even with optimal surgical nerve reconstruction and regeneration of the sensory and motor nerves, synkinesis can be so pronounced that use of the hand is impaired. A good surgical and anatomical result can thus be com-

pletely unsatisfactory as far as function is concerned.

The sensory system is naturally not exempted from such aberrant regeneration. The ability to perceive an object with eyes closed is based on a chain of events called stereognosis. This starts with taking in and transmitting information to the brain from sensory and pressure receptors in the skin, where pattern recognition is carried out on the basis of a screening process. If the skin receptors are incorrectly reinnervated after a nerve lesion, inadequate regions of the sensory cortex are addressed and patterns arise that cannot be perceived by any familiar screen. Everyday activities such as buttoning a shirt can become problematic. Visual input can lessen the sensory deficit, but not overcome the motor effects of synkinesis. Sandpaper is used to test the quality of sensory regeneration of the fingers. Sandpaper with various degrees of coarseness is placed before the patient. He must arrange the sandpaper from fine to coarse without visual input. If reinnervation is satisfactory, he is more successful than if reinnervation is defective with a resulting loss of two-point discrimination.

Even intensive practice makes it possible to re-learn how to process motor and sensory information only to a limited extent. It was initially believed that the neuron system was in its fixed layout, difficult to modify, soon after birth. However, recent studies show that there is central plasticity and thus potential for new synaptic transmissions.

In contrast to the nervous system, the muscle is much more flexible and can adapt to any task by changing its biochemical structure. One result of denervation is the loss of contractile protein and thus of volume in the muscle fiber. The fiber can remain many years in the severely atrophied condition that develops within a few months. The muscle cells go into a condition of rest and are embedded in the connective tissue of the original muscle, which does not get smaller, but tends to grow. After reinnervation the muscle cell then receives the necessary growth stimulus and can regenerate completely within a few weeks. However, the nerve forms the muscle cell according to its needs, regardless of whether such aberrant regeneration of a muscle fiber would be useful in this place or not. These processes show that restoration of the musculature is the least critical part of a nerve lesion. Regeneration of the musculature depends solely on reinnervation, even after years. However, a satisfactory functional result can be achieved only if the connective tissue remains supple. Often scarring and hardening occur that adversely affect function. Such damage is promoted by incorrect and painful physiotherapy that injures the connective tissue and stimulates proliferation and hardening (see Sect. 10.1.3).

1.8   Sequelae of nerve injuries on the sensory system and pain

Damage to the peripheral sensory neuron can, as in motor nerves, be caused by mechanical, infectious, metabolic, or toxic noxa. The damage can involve mainly the sheath structures (Figure 6) including the myelin layer that is so essential for electrical conduction. If only the axon is damaged, it is called an axonopathy, which causes the disorders described in detail in Chap. 8. Traumatic lesions usually lead to an interruption of the neuron's continuity. Figure 7a shows various kinds of lesions of the sensory neuron. Image 2 in Figure 7a shows a lesion in the proximal segment of the nerve between the spinal ganglion and the synapse to the dorsal horn. In a lesion such as this, the peripheral segment of the nerve remains supplied by the cell body of the neuron in the ganglion. The peripheral axon segment can thus survive the damage. As the axon is still present, sensory nerve conduction velocity can continue to be recorded. Clinically of course, there is a loss of sensitivity in the supplied area due to the interruption. The electrophysiological effect is that no impulses can be conducted to the dorsal horn via the afferents. This results in deafferentation. In such a situation there are considerable changes in the neuron bundle of the dorsal horn. The transmission of information from the periphery can be massively influenced by a reconfiguration of the synapses, and degeneration of receptors up to irreversible morphological reorganization is possible. This explains why in the anesthetic area of a root avulsion in a plexus lesion of the arm, severe pain can be felt spontaneously or from a slight touch.

Deafferentation also develops when unmyelinated neurons (C fibers) are damaged, especially if they are axotomized. Due to the degeneration and lack of reinnervation of the original target tissue, many neurons die off completely. This causes dorsal horn neurons to be partially denervated. Neurons with myelinated axons (A-β and A-δ fibers) are more resistant in this respect. They even have a tendency to form new synapses toward the center via sprouting. The neurons denervated of C fibers are potential innervation partners. This process can lead to allodynia from A-β fibers (Figure 81). Mechanisms for generating pain are described in detail in Chapter 7.

If the continuity of the sensory nerves is maintained but the structures important for electrical conduction are damaged, conduction can be impaired by functional damage to the axon. Such impairment of electrical conductivity can be deter-

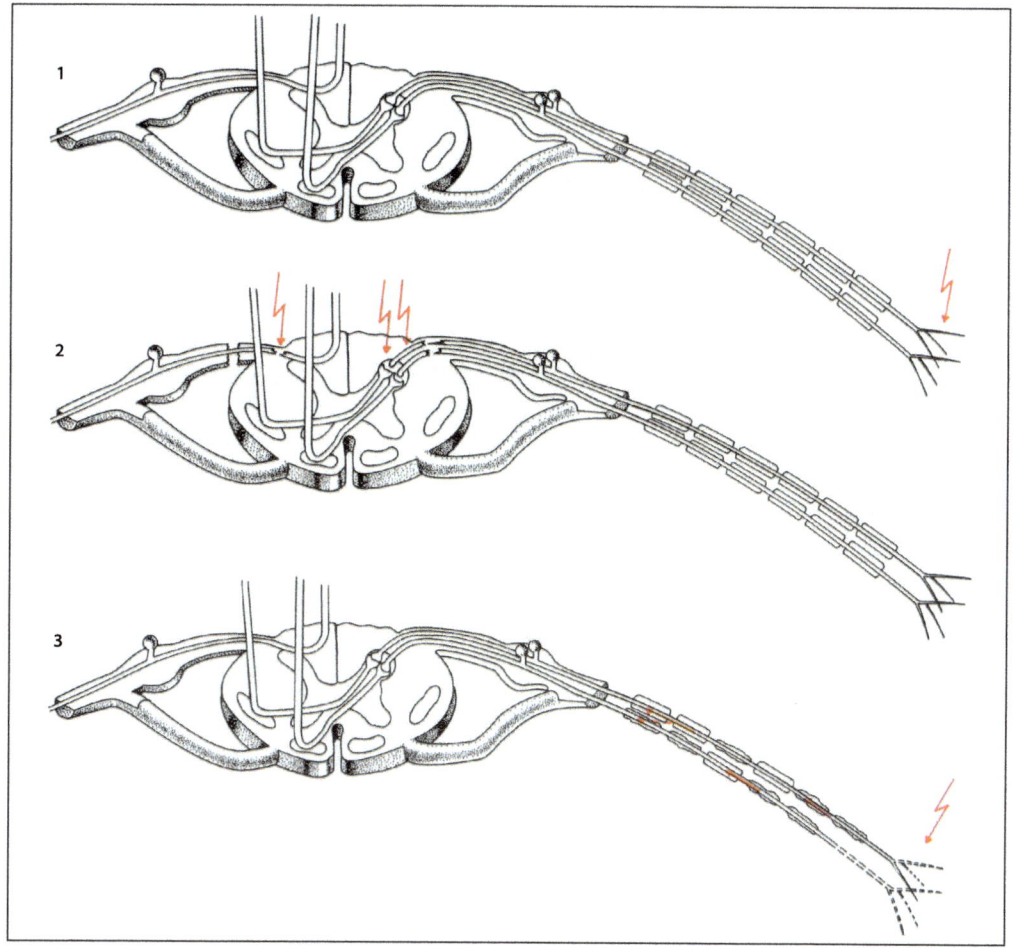

Figure 7a Sensory nerve lesion and neurogenic pain. Physiologically, impulse series that contribute to feeling pain are conducted either ipsilaterally in the dorsal funiculus toward the center or, after crossing over to the opposite side, in the contralateral spinothalamic tract. 1 Nociceptive chemical substances can activate nociceptors in tissue. 2 Root avulsion results in deafferentation of the dorsal horn neurons and neuropathic pain. 3 Lesions on the peripheral nerve allow the development of ectopic impulse generators and cross-talk between individual nerve fibers and thus neuropathic afferent activity

mined by measuring the sensory nerve conduction velocity. But this is limited to the situation involving axons with a thicker myelin. In some polyneuropathies, especially diabetic neuropathy, only thinly myelinated axons – known as small fibers – whose nerve conduction velocity cannot be measured are initially affected. In such neuropathies, patients complain continuously of severe burning pain, especially in their feet. This phenomenon can be explained by different pathological mechanisms that are described in more detail in Sect. 6.2. In peripheral nerves, the damaged myelin sheath cannot suppress cross-talk, the interference from impulses from adjacent nerve fibers. In addition, the membrane of the nerve fiber can become so electrically unstable that spontaneous impulse generation occurs. Ectopic impulse generation leads to pathological information generation (Figure 77) that is conducted toward the central nervous system.

Examination process

2

Patient descriptions of damage to the peripheral nerve usually include paresthesias such as tingling, prickling, burning, pins and needles, numbness, or poor circulation up to severe pain. If the patient mentions muscle paralysis, you need to inquire precisely what the patient means by this. Impairment caused by sensory disorders is often described as paralysis because the patient is not able to distinguish between the individual limitations and their consequences. With slowly progressive paresis that occurs with polyneuropathies or motor neuron disorders, the patient usually notices only muscle loss, lack of motor coordination, or twitching before muscle weakness is manifested. Occasionally, changes in sweating or a change in the color or texture of the skin are reported. All of these are symptoms that can be caused by irritation or deficits of sensory, motor, or vegetative nerve fibers. Only the targeted patient history – including an analysis of what the patient means by "numb" or "paralyzed"

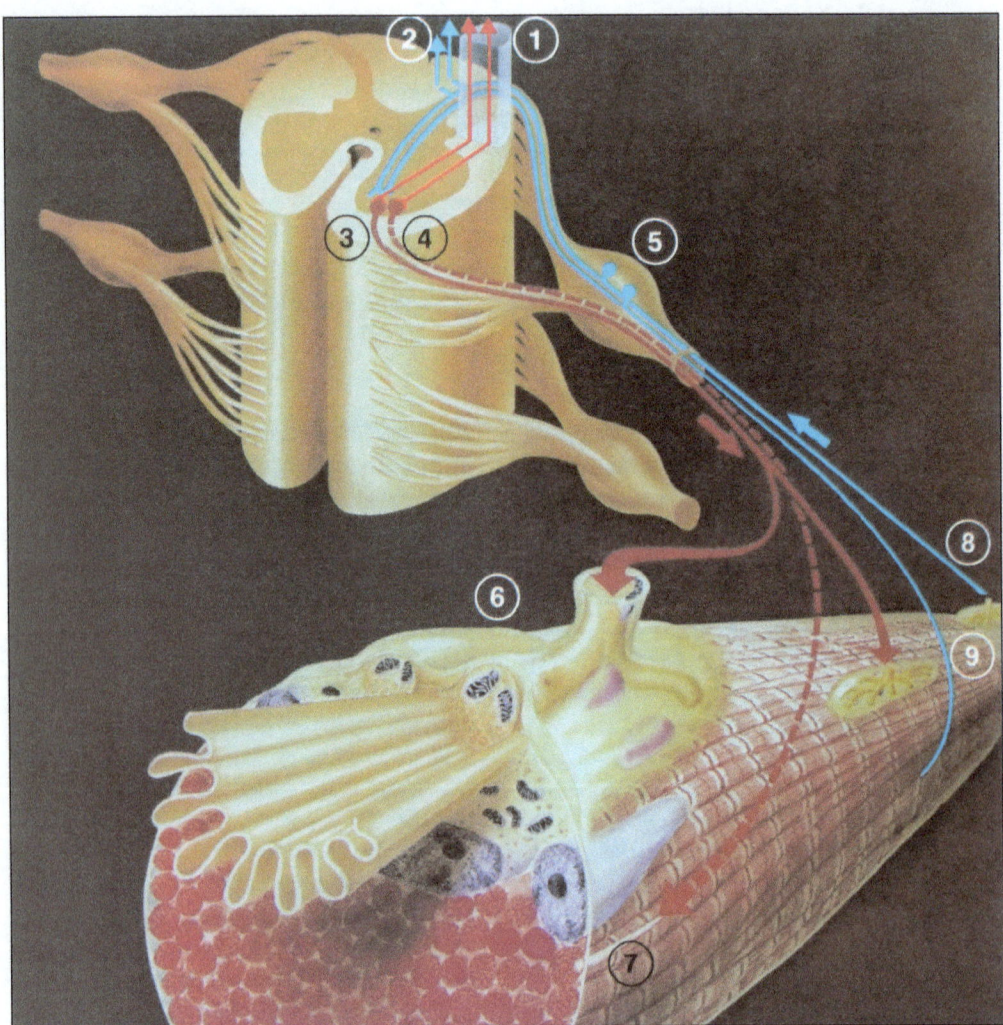

Figure 8 Spinal and peripheral structures for motor control. 1 Pyramidal tract, 2 Dorsal funiculus, 3 Ventral horn cell with axon to the muscle, 4 Gamma motor neuron to the muscle spindle, 5 Spinal ganglion, 6 Motor end plate, 7 Muscle spindle, 8 Tendon spindle, 9 Tendon

Table 2 Localization levels of paralysis	
A	**Central** (stroke, mass in the spinal canal)
B	**Peripheral** in the motor neuron (amyotrophic lateral sclerosis, spinal muscle atrophy) at the nerve root in the plexus at the peripheral nerve at the neuromuscular synapse (myasthenia gravis, poisoning, e.g., acethylcholinesterase inhibitors) due to metabolic disorders (hyper- or hypokalemia, congenital muscle enzyme deficits) due to mechanical disorders (torn muscle or ligament, bone fracture, or joint damage)
C	**Functional** (psychogenic or psychological disorder)

Table 3 Quantification of muscle strength (M)	
M 0	No palpable or visible contraction and no movement attributable to the muscle
M 1	Flicker or trace of contractions No voluntary movement
M 2	Feeble movement but not against resistance or gravity
M 3	Active movement against gravity but not against added resistance
M 4	Movement against gravity and strong resistance but can be overcome by examiner
M 5	Normal strength and range of movement

and where exactly this feeling is located – coupled with a complete examination leads to discovering the origin and localization of damage. The examination must consider whether the symptoms indicate damage in the central or peripheral nervous system or whether they should be assessed as a functional disorder (Table 2).

A systematic examination process is the best guarantee for a complete, thorough physical examination. The test of motor function and reflexes should be followed by an examination of sensitivity and vegetative functions.

2.1 Motor function

The active or passive movement of segments of extremities in the natural range of movement of their joints shows initially whether paralysis is caused by nerve damage or defects in osseous, connective tissue, or joint structures. After this, a targeted test of individual muscles is necessary to obtain an overall picture of the damage pattern. Paresis can be determined by the innervation pattern of a peripheral nerve or indicate a lesion of the nerve root. Sym-

metrical paresis that does not follow an innervation pattern can be caused by a polyneuropathy or also by a primary muscle disorder. Therefore, in the examination not only the deficit pattern must be determined, but also the degree of paresis of a muscle. This determination is made depending on what resistance can be overcome in the muscle test and is made using defined muscle strength degrees (M) (Table 3). This international scale according to the guidelines of the Medical Research Council in London also facilitates the assessment of a paralyzed muscle during the course of the reinnervation phase, which usually takes several months.

2.1.1 Muscle tests of the upper extremity

Figure 9A–I shows the function tests for various muscles of the upper extremity. In the parentheses following the muscle are the nerves and the segmental influx that innervates it. The direction in which the muscle must take the extremity in the test is indicated with red arrows.

2

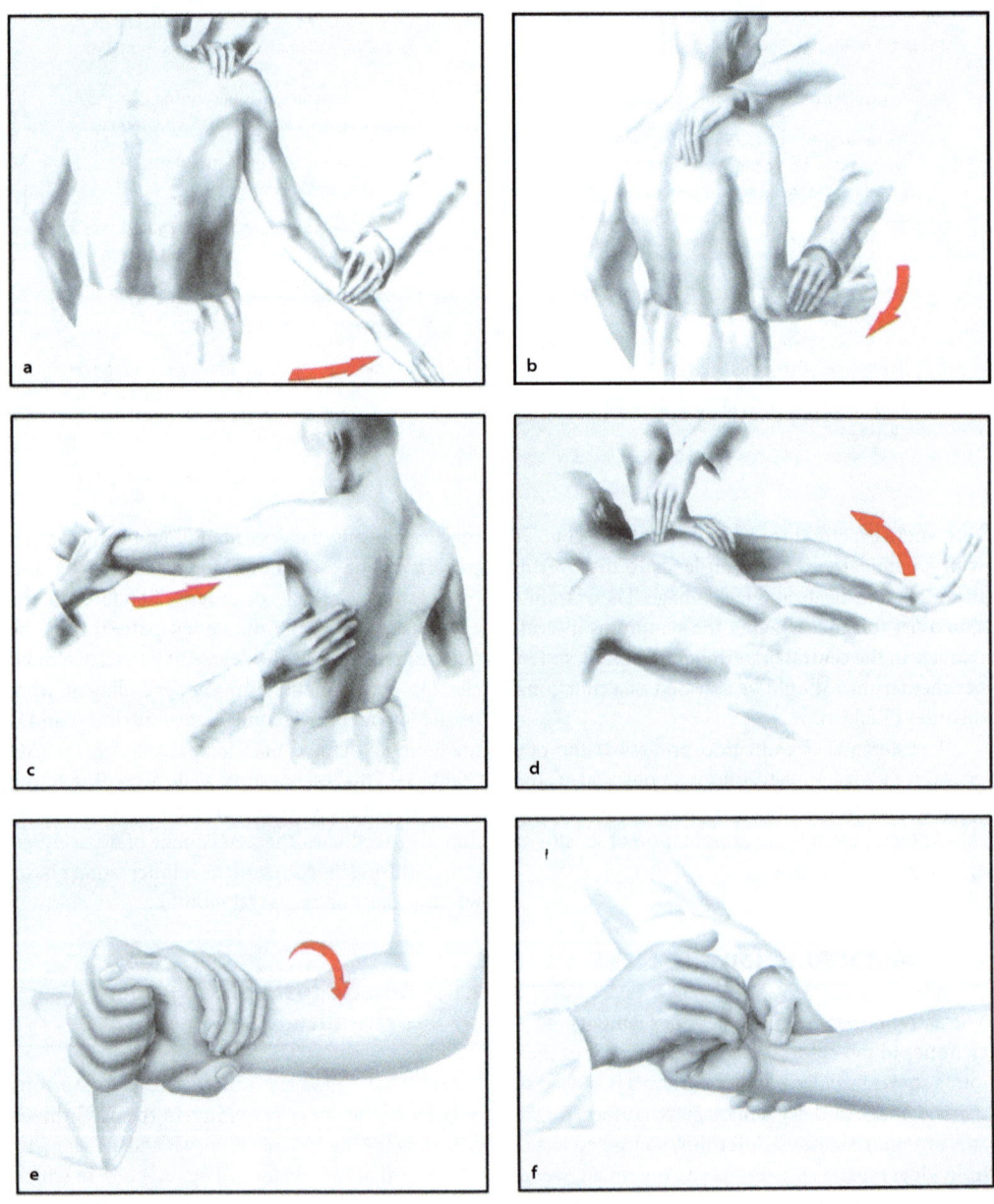

Figure 9a–f

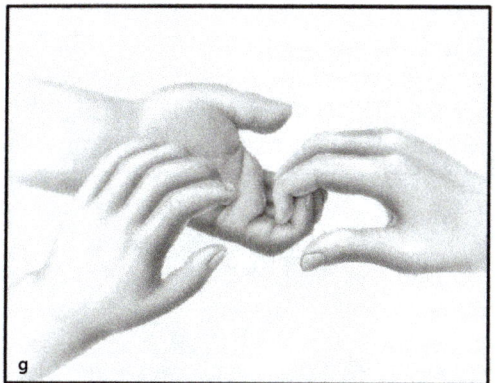

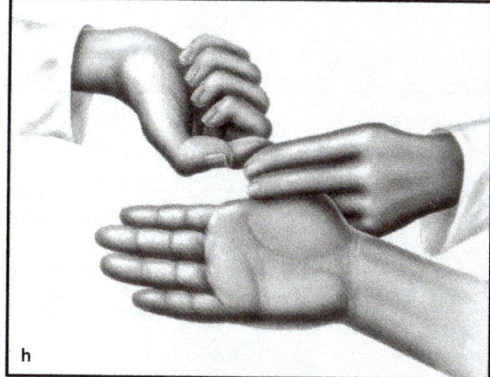

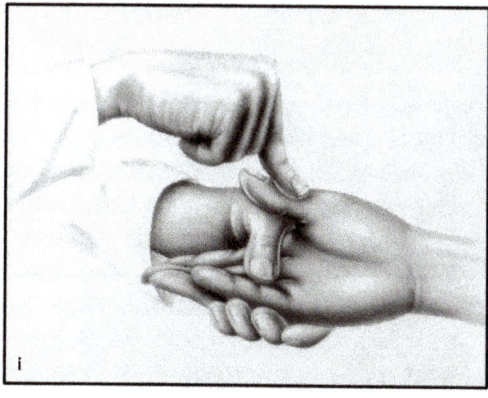

Figure 9 Manual muscle testing – upper extremity. **a** Function test of the suprascapular muscle (suprascapular nerve, C5–C6). Abduction of the arm shows weakness, especially for the first 15°. **b** In the function test of the infraspinatus muscle (suprascapular nerve, C5–C6) and the teres minor muscle (axillary nerve, C4–C5), the upper arm placed against the upper body must conduct an outward rotation. A deficit of the infraspinatus muscle results in a clear weakness in outer rotation in the shoulder. **c** The serratus anterior muscle (long thoracic nerve, C5–C6) attaches the shoulder blade to the thorax. When the extended arm is lifted by a paretic muscle, it cannot be lifted above a horizontal line. In this maneuver, and even more in push-ups, sometimes even while standing, the scapula extends beyond the thorax like a wing. When the serratus anterior is weak, the lower angle of the scapula will wing out and be displaced toward the center and upward. The lack of fixation of the medial edge of the shoulder blade leads to scapula alata. This phenomenon can also be found in muscular diseases, in particular muscular dystrophy. **d** The function test of the rhomboid muscles (dorsal scapula nerve, C4–C5) is best performed on a patient lying down. Atrophy and paresis are apparent when the patient presses his shoulders backwards and at the same time tries to cross his outstretched arms behind his back. **e** Pronation of the forearm with the elbow bent is carried out by the pronator teres muscle and the pronator quadratus muscle (median nerve). The brachioradial muscle (radial nerve, C5–C6) is responsible for supination. **f** In the function test of the flexor carpi radialis muscle (median nerve, C6–C7) the examiner holds his thumb on the tendon of the muscle. The tendons of the long palmar muscle (median nerve, C7–C8) can be seen by flexing the wrist. The flexor carpi ulnaris muscle (ulnaris nerve, C8–T1) flexes the hand and allows it to carry out a slight ulnar abduction. **g** The flexor digitorum superficialis muscle (median nerve, C7–C8) flexes the fingers in the 1st phalangeal joint with the wrist fixated. **h** Function test of the flexor pollicis longus muscle (median nerve, C6–C8). The tip of the thumb is flexed. **i** Function test of the abductor pollicis brevis muscle (median nerve, C7–T1). The metacarpus is fixated, the muscle lifts the thumb at a right angle from the plane of the hand

2.1.2 Muscle tests of the lower extremity

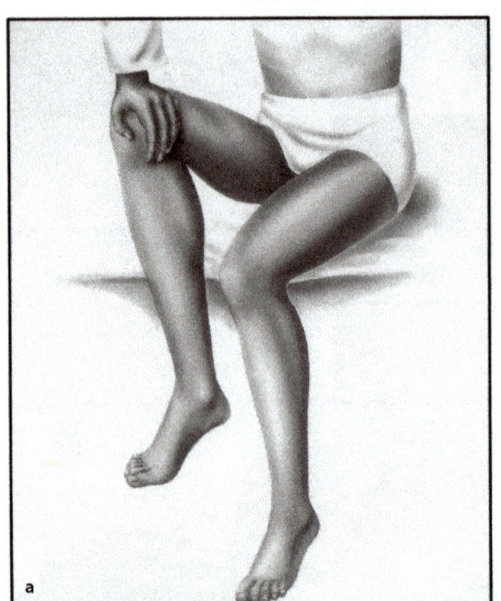
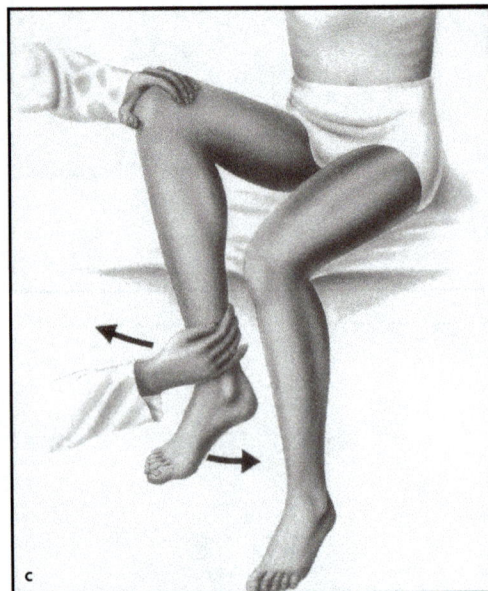
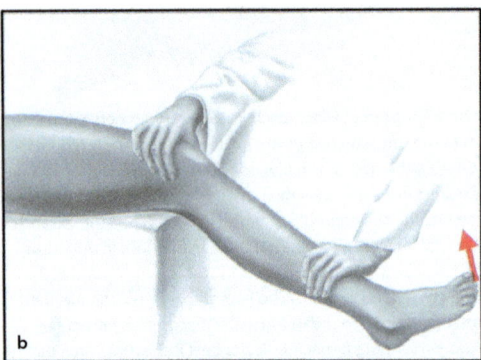

Figure 10 Manual muscle testing – lower extremity. **a** The iliopsoas muscle (femoral nerve, L1–L3) is the strongest flexor in the hip joint. To test it, the sitting patient must bend his thighs against the examiner's resistance. **b** The quadriceps muscle (femoral nerve, L2–L4) is the strongest extensor of the lower leg. The sitting patient lets the lower leg hang down from the examination couch from the back of the knee. If there is severe muscle paresis, the lower leg cannot be lifted against gravity. **c** Muscles innervated by the sciatic nerve are involved in flexing the lower leg. These are the biceps femoris muscle (L5–S1), which with its tendons forms the lateral edge of the back of the knee as well as the semimembranosus muscle (L5–S2) and the semitendinosus muscle (L5–S2), which with their tendons form the medial edge of the back of the knee. The sartorius muscle (femoral nerve, L2–L3) flexes the lower leg medially to the lotus position. The examiner provides resistance against the muscle's direction of motion

2.2   Reflexes

The determination whether a reflex is normal, weak, or enhanced should be made only after repeating the test several times and always in comparison with the contralateral side and in association with the other muscle reflexes (Figures 8, 11). There is a wide variation in the normal amplitude of response. Moreover, there is variation in response in the same person in both health and disease from time to time. In addition to the proprioceptive reflexes that are monosynaptic with stretching of the muscle spindles and contraction of the same muscle (Figure 10), there are polysynaptic reflexes such as the Babinski reflex (Figure 12, F), in which a sensory stimulus on the sole of the foot triggers plantar flexion of the toes. The triggering site and the target organ are different. Damage to the central neuron causes enhancement of the monosynaptic and weakening of the polysynaptic reflexes. Loss or weakness of both reflexes indicates damage to the peripheral neuron. Knowledge of these guidelines can be used to locate the damage via reflex disorders. For example, the cause of a foot extensor paresis if the Achilles tendon reflex is normal is probably due to a peripheral lesion in the region of the peroneal nerve. If the reflex is absent, root damage at L5 is likely, and an enhanced reflex leaves no doubt of central damage.

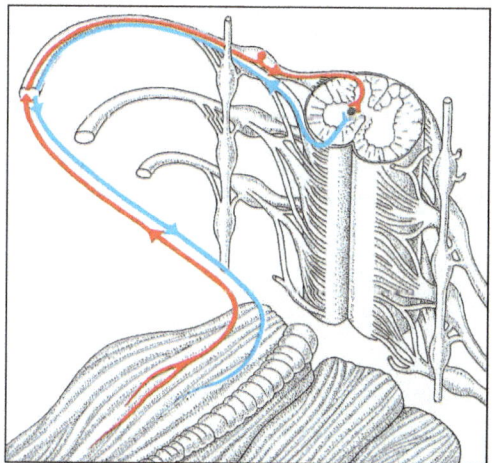

Figure 11 Monosynaptic reflex arc

2.2.1   Examination of reflexes

A skeletal muscle responds to a sudden stretch with a reflex contraction mediated by a simple and usually monosynaptic reflex arc. The afferent side of the arc begins with the muscle stretch receptor (Figure 8, Nos. 7, 8), the cell bodies of which are in the dorsal root ganglia (Figure 8, No. 5). Intramedullary fibers of these cells synapse on the motor neuron in the ventral horn of the spinal cord. The efferent side of the arc is the motor neuron with its axon and terminal structures that innervate the muscle.

Central influences modify the responsiveness of the motor neuron as well as the complex stretch receptors in the muscle. The response of the reflex arc depends on its own integrity and on the state of the central nervous system. Tests of the clinically most important monosynaptic reflexes of the upper and lower extremity are shown in Figure 12.

2.3   Sensitivity

The topographic distribution of a disorder allows a distinction to be made between a peripheral and a central lesion. Normally, the peripheral sensory lesions can be classified as the familiar picture of striated, segmental, or typical macular innervation patterns of peripheral nerve lesions, as presented in Figure 13a and b for the front and back side of a human. Sensory disturbances can be further differentiated into plus and minus signs (Table 4). Thus for example, hypoalgesia is a minus (–) sign and paresthesia a plus (+) sign. Usually the testing for sensory qualities is performed qualitatively. In the domain of pain research, a standardized protocol is used to characterize the somatosensory phenotype of patients with neuropathic pain – quantitative sensory testing (QST). QST measures the detection threshold of accurately calibrated sensory stimuli. Vibratory, thermal, or painful stimuli are often chosen because they relate to distinct neuroanatomic pathways with discrete fiber populations. However, natural stimuli rarely activate single types of receptors but rather activate different combinations of receptors. This protocol helps to identify symptom patterns that can be associated with pain mechanisms involved in neuropathic pain. As therapeutic

2

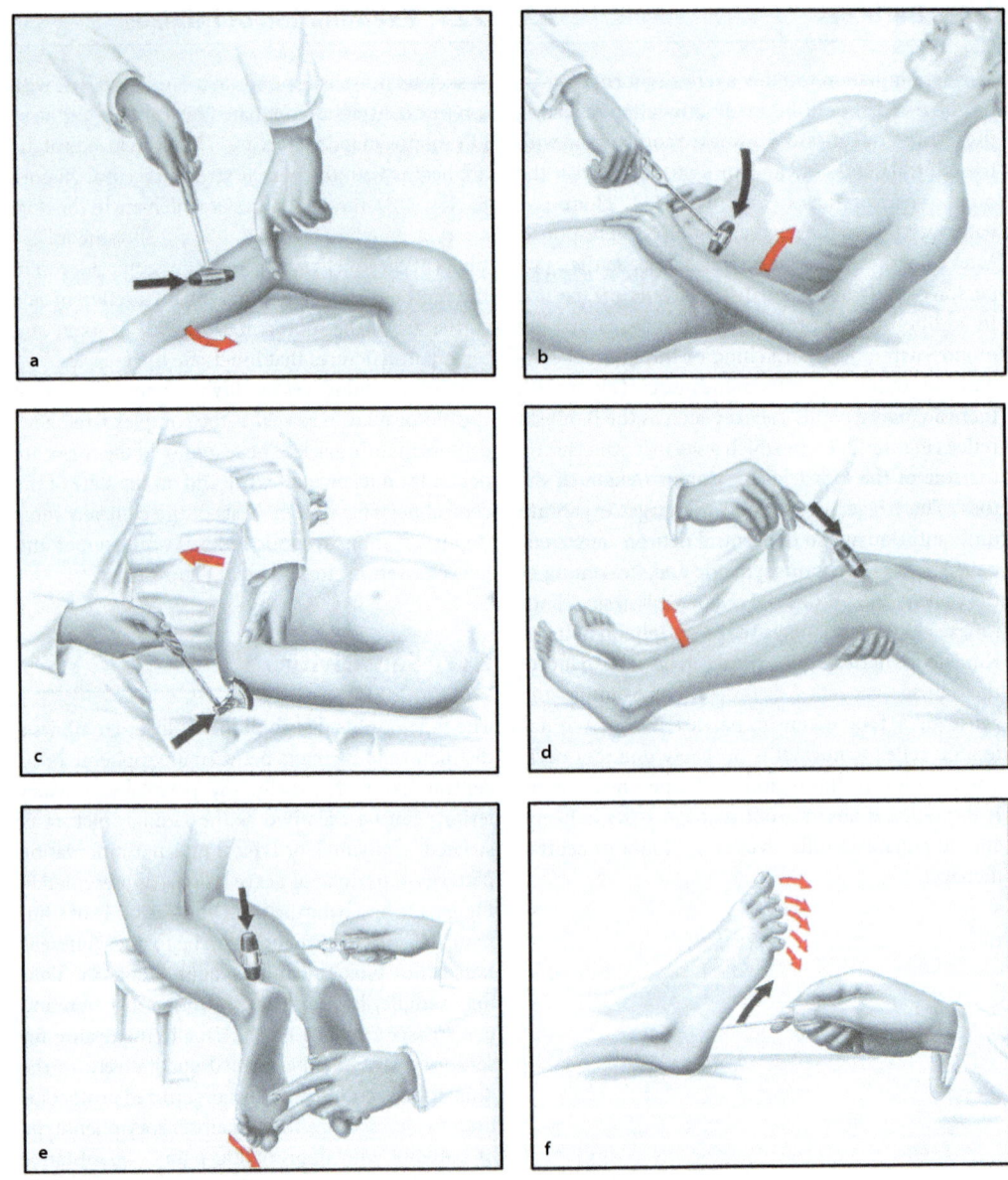

◀ **Figure 12** Reflexes. **a** Biceps reflex: The reflex extends via the cervical segment C5–6 and the musculocutaneous nerve. A sharp tap on your own finger placed in the antecubital fossa of the slightly bent arm elicits a contraction of t he biceps muscle. **b** Radioperiosteal reflex (also known as brachioradial reflex): Its reflex arc extends to the radial nerve and musculocutaneous nerve via segment C5–6. Tapping the supinated, slightly flexed forearm in the lower third of the radius elicits a contraction of the biceps and brachioradial muscles. **c** Triceps reflex: The monosynaptic circuit is in the C7–8 segment with the radial nerve as the peripheral reflex pathway. The reflex hammer should be tapped directly on the triceps tendon above the olecranon. This causes the slightly bent forearm to extend slightly due to a contraction of the triceps muscle. **d** Patellar reflex: The monosynaptic reflex of the quadriceps muscle is mediated via the femoral nerve and the L3–4 segment. In supine position, the relaxed legs are bent by raising the backs of the knees. Tapping on the patellar tendon leads to movement through contraction of the quadriceps muscle. **e** Achilles reflex: As a monosynaptic reflex of the triceps surae muscle, it is used in S1–2 segment diagnostics, where the tibialis nerve forms the afferent and efferent reflex arc. It can be elicited on the lying or kneeling patient by tapping the Achilles tendon. Plantar flexion is triggered in the foot held at a right angle. **f** Babinski reflex: This polysynaptic reflex is elicited by stroking the sole of the foot and the response is a plantar flexion of all toes. If there is a lesion of the pyramidal tract, the response is different – the toes spread out like a fan; the large toe is extended dorsally

regimens are now available for some of these mechanisms, the exact evaluation helps to choose a mechanism-based therapy. Since it is also possible that the paingenerating mechanism and the symptoms change during the course of the disease, the therapeutic regimen must be adaptable.

2.3.1 Methods of testing sensitivity

The senses of touch and pain are tested. The examination procedure should be first explained and begun in a healthy area so the patient knows what is being tested. Then the testing is done in the affected area, ideally in comparison with the comparable region on the contralateral side, as sensory nerve supply is not the same in all areas of the body. If the margins of a disorder are then to be determined, the region is tested from the healthy area to the deficient area and then in the other direction for confirmation.

Tactile perception

Tactile esthesia is best tested with a cotton swab that creates equal, reproducible examination conditions when applied with light pressure. Too much pressure does not test the function of the nerves in the skin, but rather subcutaneous pressure receptors. In polyneuropathies, in particular diabetic polyneuropathies, these two modalities can be deficient to a different extent. In comparison with a healthy skin area, a stimulus can be perceived as weak (hypoesthesia), or enhanced (hyperesthesia). Spontaneously occurring sensations such as "pins and needles," "tingling," or "electricity," are considered

plus symptoms and are described as paresthesia. But if they are persistent and clearly unpleasant, they are called dysesthesia. Esthesia and algesia differ from each other to varying degrees between the individual segments and are therefore of diagnostic interest. Skin areas in which the same sensory quality of several nerves is registered are called intermediate zones; individually supplied areas are called autonomous zones. The autonomous zones are useful, typical, and specific deficit patterns for diagnosing peripheral nerve damage (Figure 14 A–E).

Pain perception

Algesia is tested with a pain stimulus (Figure 15). A sharp pin can be used. It should be used lightly and consistently to elicit just a sensation of sharpness and pain. Start in an unaffected area and then go to the affected area. Use the unaffected area as reference for reorientation to define the borders of impairment. Oversensitivity to a stimulus normally perceived as painful is called hyperalgesia; undersensitivity is hypoalgesia. In sensory deficits caused by peripheral nerve damage, hyperalgesia can repeatedly occur in the area supplied by the nerve and a painful sensation can be elicited from a stimulus that is not painful at all on undamaged skin. In postherpetic neuralgia we often have the phenomenon of anesthesia to pin prick but light mechanical irritation with a cotton swab can elicit intense pain, which then would be called dynamic mechanical allodynia. In pain there are independent pathophysiological mechanisms in both the peripheral and central nervous systems that are responsible for sensory symptoms as well as spontaneous and evoked

2

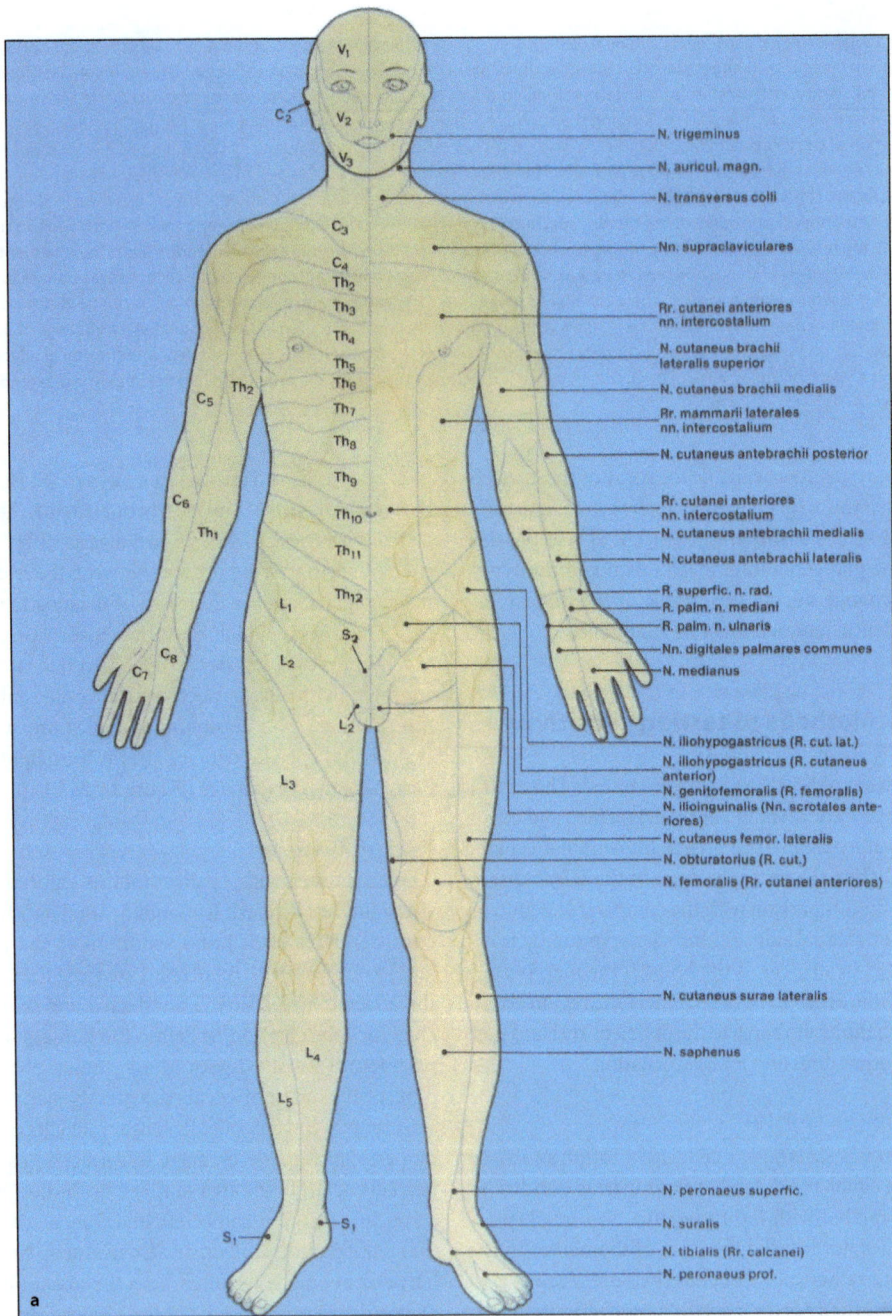

Figure 13 Diagram (**a**, front; **b**, back) of the innervation pattern of spinal segments and peripheral nerves. For symmetric stocking-glove paresthesia, additional examinations including reflex and muscle status and neurographic measurements are necessary. Only these can distinguish between a spinal cord transection with enhanced reflexes, a polyneuropathy with weakened reflexes and slower nerve conduction velocity, or polyradiculitis with weakened reflexes and normal nerve conduction velocity. We see again and again that the patient's indication of negative signs such as numb feet is frequently assessed as peripheral circulation disorder and not acted on. Easily palpable foot pulses help to rule out circulation problems and classify the symptoms as signs of polyneuropathy

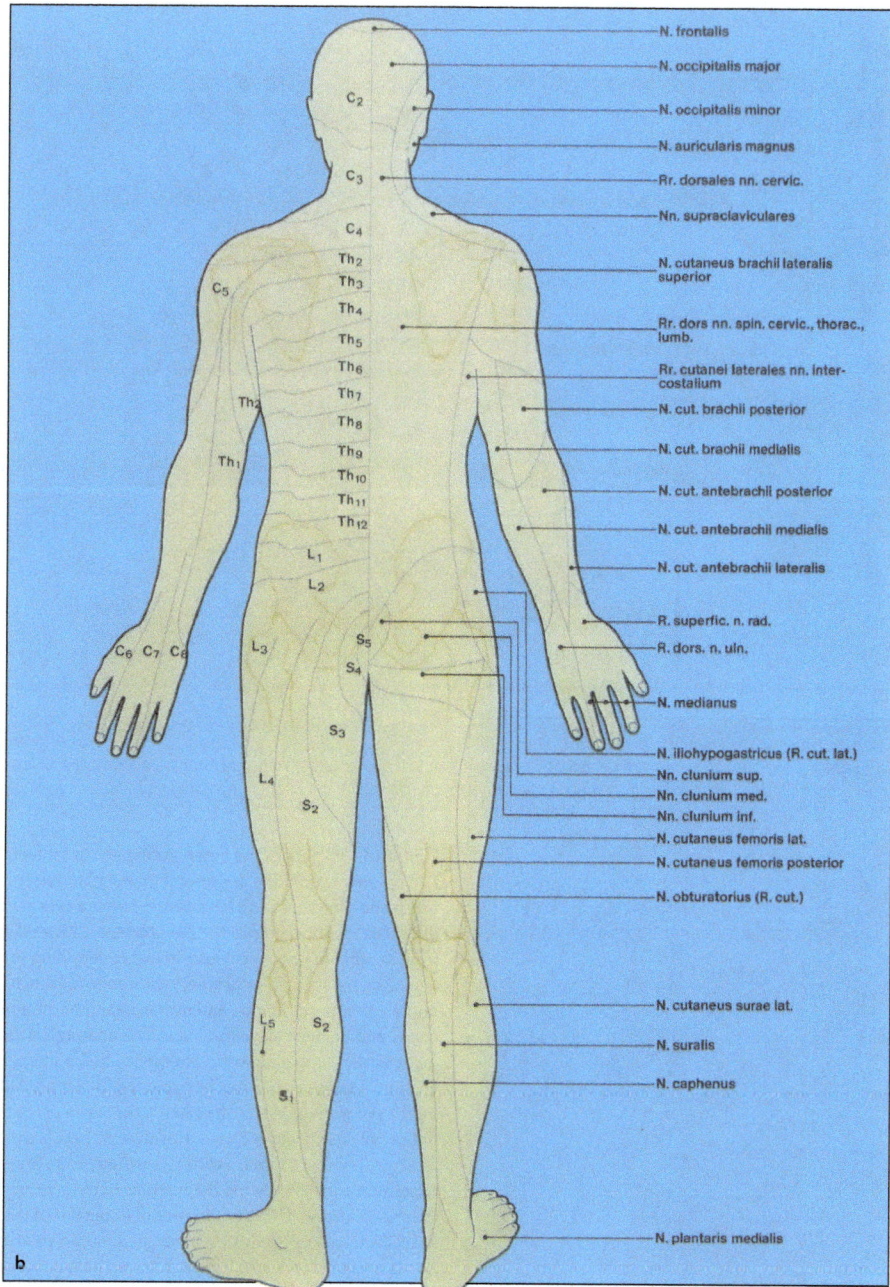

N. frontalis
N. occipitalis major
N. occipitalis minor
N. auricularis magnus
Rr. dorsales nn. cervic.
Nn. supraclaviculares
N. cutaneus brachii lateralis superior
Rr. dors nn. spin. cervic., thorac., lumb.
Rr. cutanei laterales nn. inter-costalium
N. cut. brachii posterior
N. cut. brachii medialis
N. cut. antebrachii posterior
N. cut. antebrachii medialis
N. cut. antebrachii lateralis
R. superfic. n. rad.
R. dors. n. uln.
N. medianus
N. iliohypogastricus (R. cut. lat.)
Nn. clunium sup.
Nn. clunium med.
Nn. clunium inf.
N. cutaneus femoris lat.
N. cutaneus femoris posterior
N. obturatorius (R. cut.)
N. cutaneus surae lat.
N. suralis
N. caphenus
N. plantaris medialis

b

Figure 13b

2

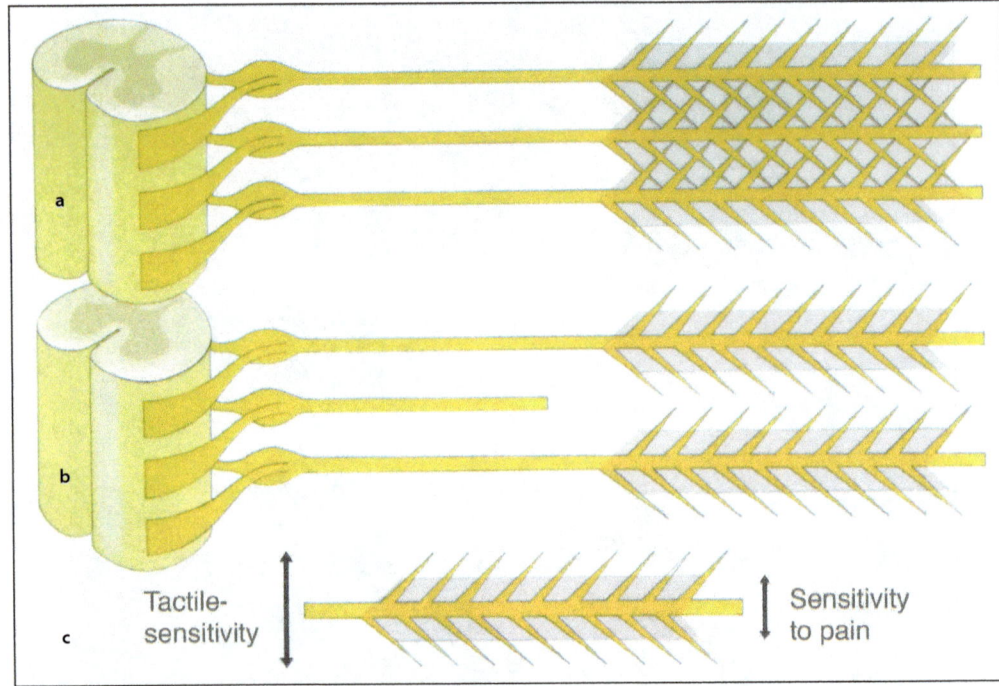

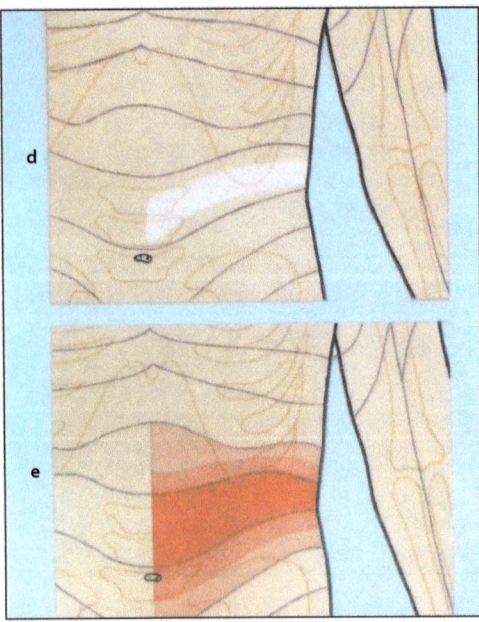

Figure 14 Diagram of the innervation pattern for tactile and pain sensation. The segmental innervation zones for touch and pain sensation are arranged on the skin at varying distances (**c**). For example, the pain zones are located next to each other in narrow circles without overlapping (**b**, gray area). The monosegmental innervation zone for touch, on the other hand, is wide and overlaps far into the zones above and below it (**a**, yellow lines). This anatomical arrangement explains the diagnostic experience that a sharply delineated analgesic zone can be expected for a monoradicular deficit syndrome. On the other hand, the sense of touch is not or only minimally affected. However, if there is a monoradicular pain syndrome, a wide dysesthetic zone often arises that extends beyond the margins of the segment. As described above, deficit syndromes of a spinal root lead to a small analgesic zone (**d**). However, if monoradicular irritation is present, for example, from referred pain or herpes zoster, the skin shows dysesthetic changes in a much broader zone (**e**)

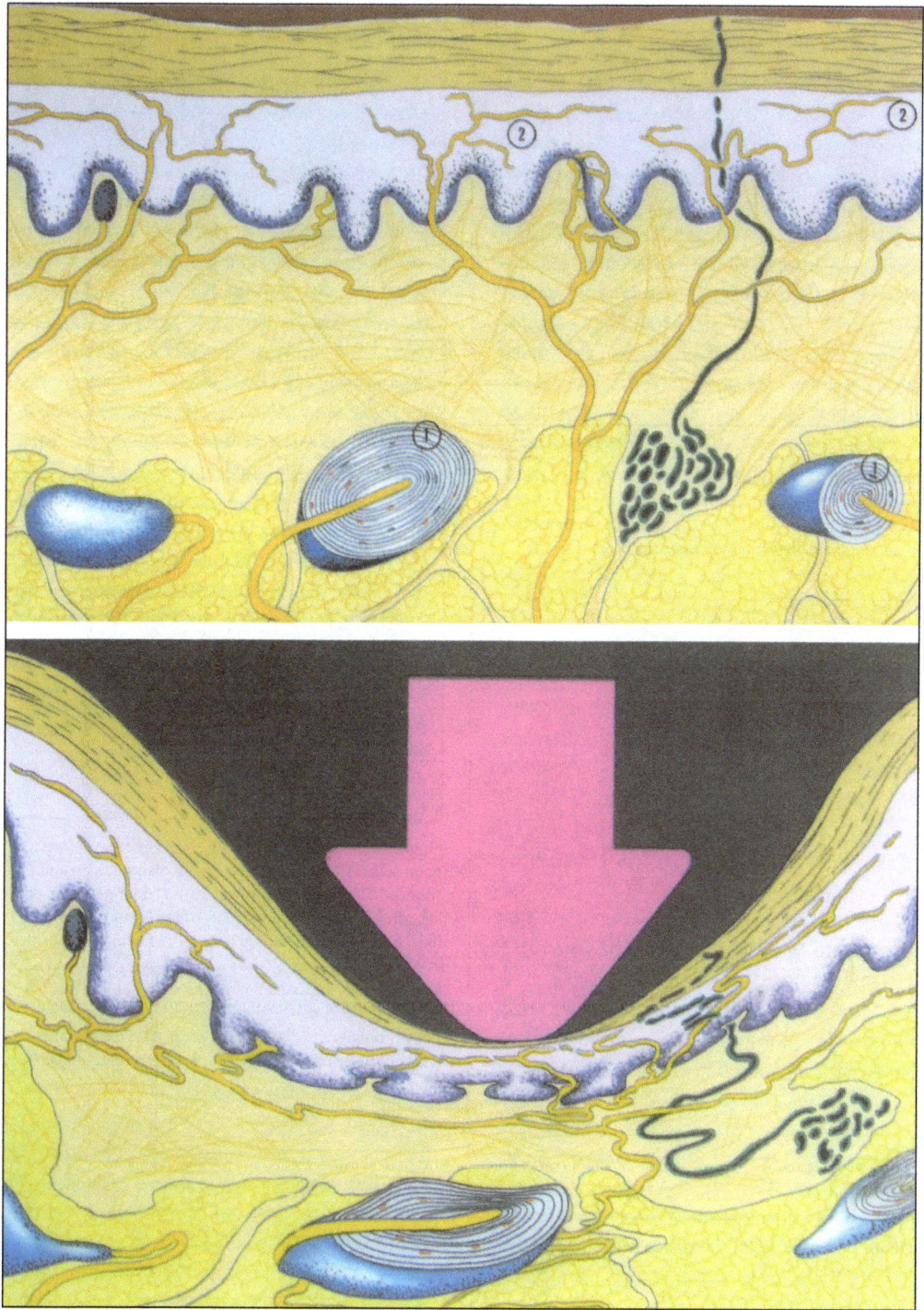

Figure 15 Effect of pressure on receptors in the skin. Mechanoreceptors (1) and nociceptors (2), which are the free nerve endings, are distributed in skin and mucosa at varying densities depending on the functional need (see also Figure 71). Receptors are deformed, here through pressure. This change generates action potentials that are conducted towards the center via the nerve

2

Figure 16 Free nerve ending. The free nerve ending is an unspecialized afferent nerve ending. These nerves are involved in detecting temperature, mechanical stimuli (such as pressure), and touch. Since they are also responsible for detecting pain, they function as nociceptors

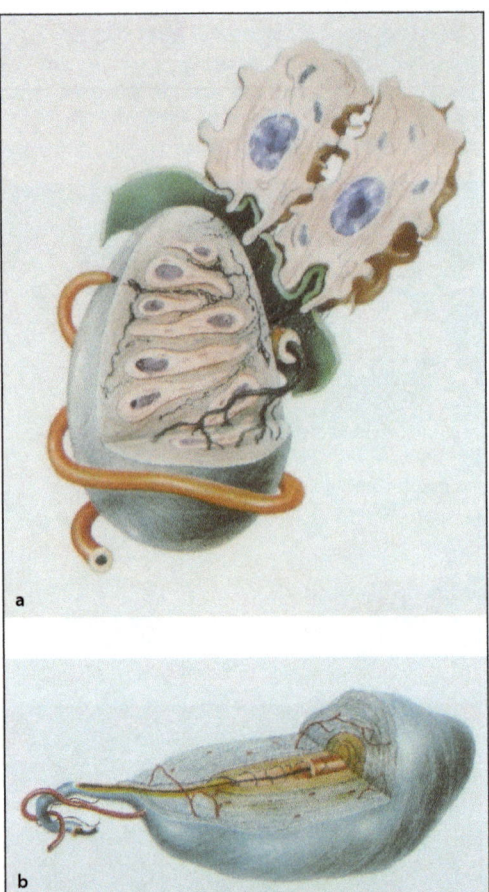

a

b

Figure 18 Receptors for pressure and vibration. **a** Meissner's corpuscle is situated in the dermis of glabrous skin and responds to vibration sensitive at 20–40 Hz range. **b** The pacinian corpuscle can be found in the deep layers of dermis in hairy and glabrous skin and responds to vibration sensitive at 150–3000 Hz range. The test for vibration sense can be performed with a Rydel-Seiffer graduated tuning fork (64 Hz, 8/8 scale). The fork has to be activated and placed over a bony prominence (ulnar styloid process, internal malleolus). It is kept there until the subject reports the end of feeling vibration

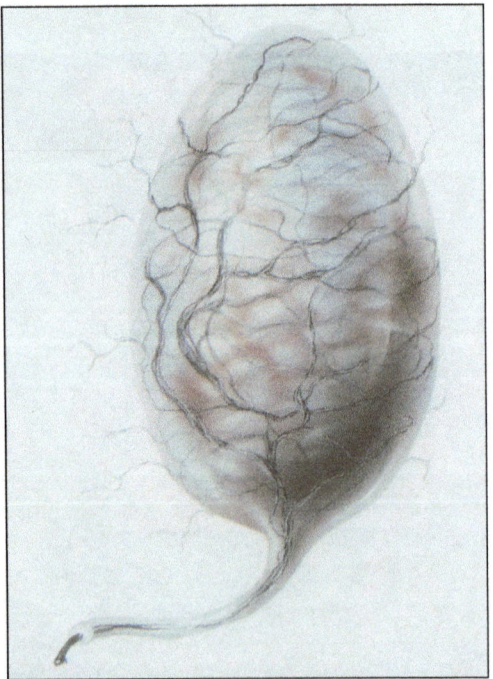

Figure 17 Krause's corpuscle is a rapidly adapting receptor responsive to pressure

pains after nerve lesions or in peripheral neuropathies. The following examples explain which mechanisms can underlie the clinical symptom.

Dynamic mechanical allodynia is an example where only the nociceptor function is selectively impaired within an allodynic situation of the skin. Pain and temperature sensation are profoundly impaired but light moving mechanical stimuli can often produce severe pain (dynamic mechanical allodynia). The reason for this lies in reorganization in the dorsal horn. C fiber degeneration induces a new anatomical connection of A-β fibers with central pain pathways (Figure 80).

Another possibility is pathological active or sensitized nociceptors that in a second step can induce secondary changes in the chain of central pain processing. Here the induced spinal cord hyperexcitability is important and in this state the input from mechanoreceptive A-β fibers (light touching) is then perceived as pain. The same mechanisms can underlie pain states with spontaneous pain, heat hyperalgesia, and static mechanical allodynia as leading symptoms (Figure 81).

In low back pain, a persistent inflammatory reaction of the nerve trunk can induce ectopic activity in primary afferent nociceptors and thus is a potential cause of spontaneous pain and allodynia. This effect is mediated by the cytokine tumor necrosis factor alpha (TNF-α) produced by activated macrophages. As the nociceptors in the free nerve endings are chemosensors, they can be activated by a wide range of inflammatory molecules. Besides protons (H+) and bradykinin, these are: postaglandin, adenosine-5-triphosphate (ATP), nerve growth factor (NGF), and serotonin. These mediators are released by epithelial cells, macrophages, and mast cells (see Figure 89).

After nerve lesion, the sympathetic nervous system might interact with afferent neurons. Activity in sympathetic fibers can induce further activity in sensitized nociceptors and, therefore, enhance pain and allodynia (sympathetically maintained pain). This pathological interaction acts via noradrenaline released from sympathetic terminals and newly expressed receptors on the afferent neuron membrane (Figure 82).

These mechanisms can operate in concert in a single disease entity (e.g., postherpetic neuralgia) and also in a single patient. Distinct pathophysiological mechanisms lead to specific sensory symptoms (e.g., dynamic mechanical allodynia, cold hyperalgesia). A thorough analysis of sensory symptoms may reveal the main underlying mechanisms that are active in a particular patient. The treatment of neuropathic pain is currently unsatisfactory (see Sect. 10.3.1). In the future, drugs will be developed that address specifically the relevant combination of mechanisms.

Thermal perception

Testing for thermoesthesia is significant only for differential diagnostic purposes. For the test, the coldness of a metallic reflex hammer and the warmth of the examiner's hand are generally sufficient as gross test stimuli. If the test of a skin zone shows that the sense of pain and temperature are lacking where the sense of touch is maintained, this indicates a dissociated sensory perception. This is indicative of a central lesion in the spinal cord and can be triggered by a spinal gliotic rod or an insult in the area supplied by the anterior spinal artery. In the spinal cord, the fibers for the sense of pain and temperature cross over to the contralateral side before the central canal (Figure 7a). Since the other qualities are conducted toward the center via the posterior tracts without crossing, they avoid any damage located at the central canal.

Spatial perception ability

The test of the ability to discriminate two points is very useful for diagnosing and controlling the progress of disorders. The distance between two tips of a compass can be varied for the test and recorded precisely in millimeters. The spatial resolution ability is exceeded when touching with two tips of the compass is perceived as only one tip. Depending on the varying density of pressure receptors in the skin, not all skin zones have the same level of discrimination. The most sensitive are the pads of the fingers, which can perceive a minimum distance of 2 mm. On the back, the minimum distance is 70 mm; on the forearm it is 40 mm. Two-point discrimination can also be tested using various grades of sandpaper. If a patient can sort sandpaper according to coarseness without visual control, a discrimination disorder can be ruled out.

2

Table 4 Bedside testing of sensory negative and positive signs in neuropathic pain

	Symptom	Definition	Test	Response
Negative (−) Signs	Hypoesthesia	Reduced perception of a non-painful touch	Stroking the skin with a cotton swab	Numbness, reduced sensitivity to touch
	Pallhypoesthesia	Reduced perception of vibration	Stimulation of a bone prominence with a 64-Hz tuning fork	Reduced perception
	Hypoalgesia	Reduced perception of a painful stimulus	Pinprick testing on the skin	Numbness
Positive (+) Signs Spontane- ous Pain	Paresthesia	Pins and needles sensation	Measure area and define intensity by visual analogue scale (VAS)	
	Lancinating pain attacks	Electric shock-like episodes	Define number of shocks per time, eliciting factors, intensity by VAS	
	Superficial pain	Sensation of pain of burning character	Measure area, intensity by VAS	

Quantitative sensory testing

QST is a tool for measuring sensory impairment for clinical and research purposes. In neuropathic pain, patients complain about positive and negative signs (Table 4). The modern concept (gain of function, loss of function) of a mechanism-based treatment of pain syndromes is based on the hypothesis that different clinical signs in neuropathic pain and symptoms reflect different underlying pathophysiological mechanisms of pain generation. To get a better insight into pain-producing mechanisms and to develop therapeutic strategies, special examination techniques such as QST are needed. A test battery has been developed for neuropathic pain. Different devices are available on the market for testing. All equipment serves the same purpose – to provide graded and reproducible stimuli. The devices are used to generate specific physical vibratory or thermal stimuli and others deliver electrical impulses at specific frequencies. To test vibration sense, a stimulator generates a designed frequency. Frequencies around 200–300 Hz are optimal because pacinian corpuscles are most sensitive to vibration in this range (Figure 18). The Peltier principle is used in devices that generate thermal stimuli. The intensity and direction of current flow controls the surface temperature of a test electrode (thermode). The patient is asked to report temperature changes and sen-

sation of pain when the thermode contacts the skin. The threshold can thus be detected for heat- and cold-associated pain. The mechanical detection threshold can be measured with a standardized set of modified von Frey hairs that exert forces upon bending. The mechanical pain threshold can be measured by using weighted pinprick stimuli with a mechanical stimulator and fixed stimulus intensities.

It should be kept in mind that quantitative sensory tests are psychophysical in nature. The sensory stimulus is an objective physical event but the response represents the subjective report from a patient. Therefore, QST results should not be the sole criteria used to diagnose pathology, as psychological factors can significantly influence sensory function perception.

2.4   Vegetative nervous system

The skin contains the diagnostically most important vegetative functions of sweat secretion, piloerection, and vasomotion, which are all controlled by the sympathicus (Figure 19).

In humans there are two types of sweating. One is for the purpose of temperature regulation, especially in case of elevated temperature. Apart from that, sweating can cause wet hands and feet, a proc-

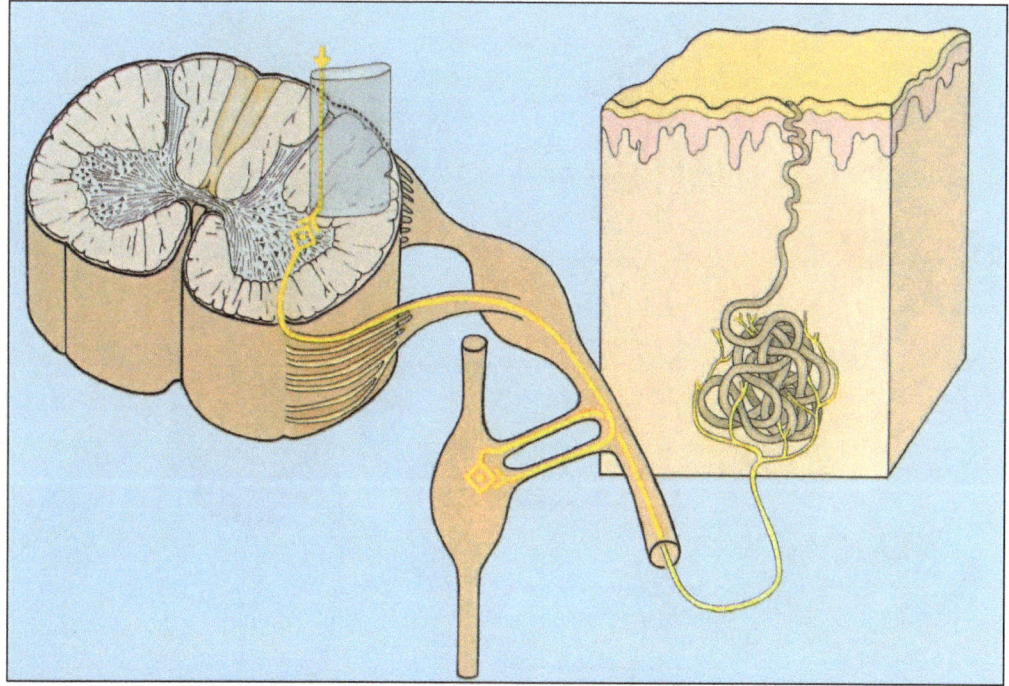

Figure 19 Eccrine sweat gland innervated by a postganglionic sympathetic fiber. Sympathetic efferents from the brain descend along the spinal cord and mediate the peripheral neuron in the lateral horn. After mediation in the sympathicus, the sweat gland is reached via the ventral root

ess that increases conductivity of skin and is apparently a phylogenetic relic.

These two kinds of sweating are tested by different means. Thermoregulatory sweating can be tested on the head, trunk, and extremities using the Minor iodine-starch test. If sweating of the palms and soles is to be tested, Moberg's ninhydrin test is used.

Any kind of sweating is possible only if the sympathetic nerve fibers are intact beyond the sympathetic trunk and in the sensory skin nerve tract. Aside from an intact peripheral tract, thermoregulatory sweating must also have functional regulatory centers in the brain and pathways to the spinal cord and sympathicus as well.

2.4.1 **Sympathetic skin innervation**

In the diagnostics of peripheral nerve lesions, innervation disorders by the vegetative system are significant, especially those caused by the sympathi-

cus. The sympathetic nerve fibers proceed to the skin along with the sensitive nerve fibers and innervate various structures there.

As already shown in the motor and sensory systems, innervation of the sympathetic system is segmental. One significant difference is that sympathetic fibers proceed from the spinal cord to the sympathicus only in the root zones C8–L2. The origin of these fibers is in the lateral horn chain of the spinal cord and they exit the spinal cord via the ventral root in the corresponding segment. However, in the sympathicus, these nerve fibers branch out so that they innervate several sympathicus ganglia above and below the exit segment. Damage in this preganglionic nerve segment can thus be easily compensated. The loss of a segmental afferent from the spinal cord is covered by the double innervation of individual sympathicus ganglia. Circumscribed sweat disorders are thus to be expected only if the postganglionic segment is damaged from the sympathicus and over the length of the peripheral nerve.

2

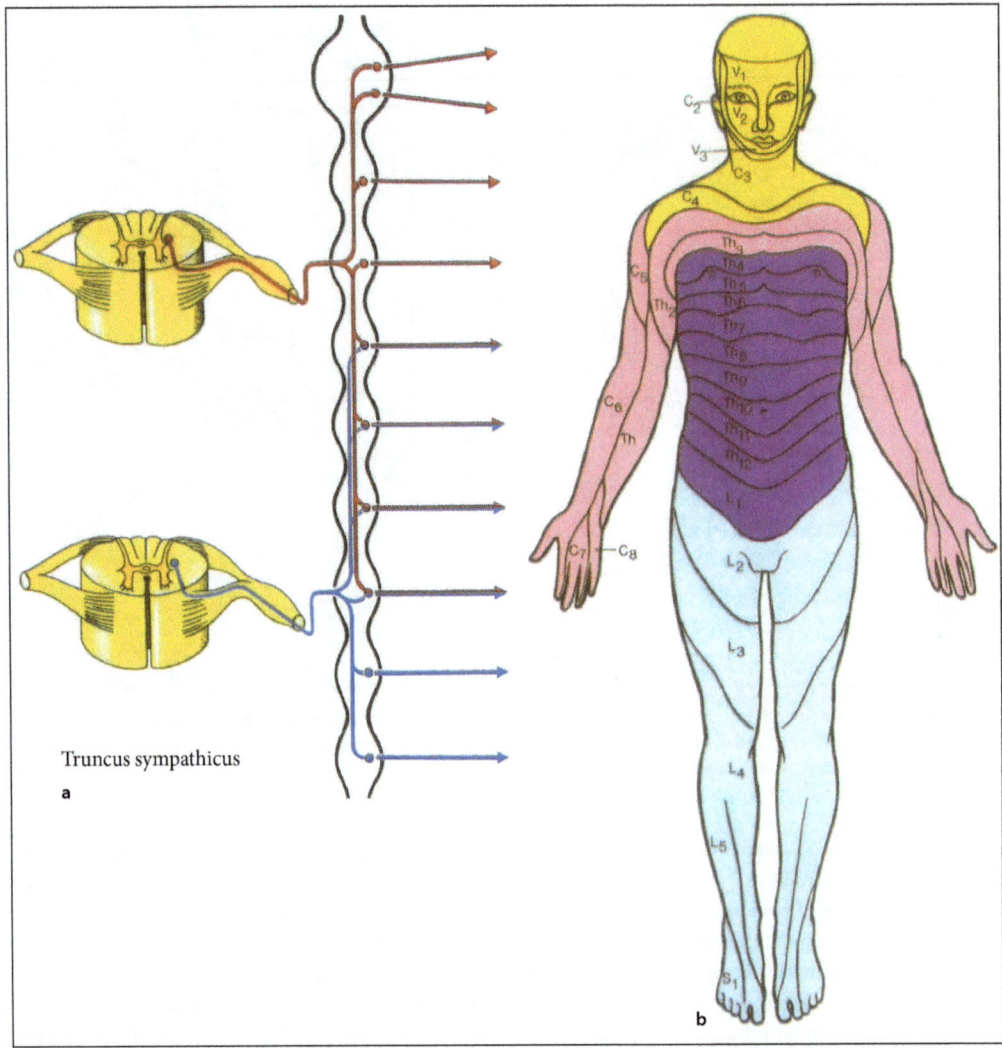

Truncus sympathicus

a

b

Figure 20 Diagram of the sympathetic innervation of the skin. **a** In the sympathetic trunk, the nerve bundle of a root makes contact with sympathicus ganglia at several levels. **b** Diagram of the metameric segmentation of the innervation of eccrine sweat glands. The yellow region (head, C2–4) is supplied from C8 to T3 to 4, the red region (C5–T3) from T5 to 6, the violet region (T4–L1) from T7 to T10, and the blue region from T11 to L2

The sympathicus runs in pairs along both sides of the spine. Three segments of the sympathicus – cervical, thoracic, and lumbar – are involved in the sympathetic supply of the skin. The cervical segment of the sympathetic trunk includes the superior, median, and inferior cervical ganglia; the latter is frequently fused with the first thoracic ganglion to the stellate ganglion.

The segmental supply for innervation of the head, arms, and upper thorax takes place via preganglionic inflow from segments C8–T3. The preganglionic sympathetic inflows from the roots T10–L2 form the sympathicus trunk, which supplies the lower trunk and legs. Figure 20 shows the metameric segmentation of sympathetic innervation.

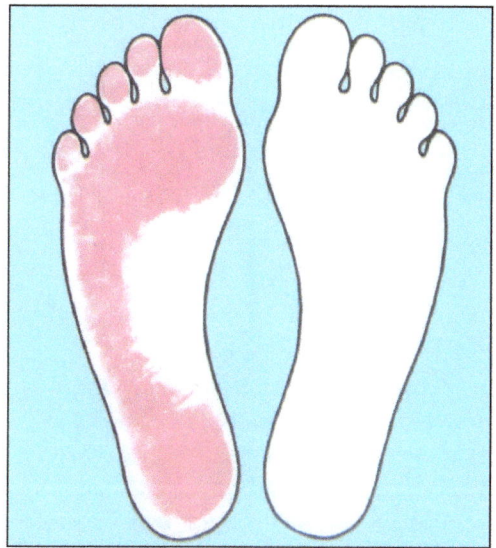

Figure 21 Ninhydrin printing test. It shows normal sudomotor activity in the right foot and missing activity in the left foot following a sciatic nerve lesion

Moberg's ninhydrin printing test

The ninhydrin printing test using Moberg's method is an easy and valid procedure for studying sudomotor activity at the palm of the hands and feet (Figure 21). Ninhydrin (triketohydrindene hydrate) is a chemical used to detect primary and secondary amines, which are contained in the sweat fluid. When reacting with these free amines, a deep blue or purple color is produced. Therefore ninhydrin is most commonly used to detect fingerprints, as amines left over from peptides and proteins sloughed off in fingerprints react with ninhydrin. Impressions of the fingertips or palm are taken on a clean piece of paper, and the finger is outlined with a pencil. White paper that has not been contaminated by previous handling is satisfactory. As a test control, a fingerprint of the examiner on the marked corner of the paper is useful. After the prints of the hands or feet are obtained, the paper is sprayed with a 1% ninhydrin solution and then dried. Good prints will appear in an hour or so, but if more rapid developing is desirable, the paper is placed in a hot oven.

Trophic disorders

The loss of sensory innervation, and closely associated with this, sympathetic innervation as well, leads to typical skin changes. The finger pads lose fullness, the skin becomes parchment thin and loses its profile. Skin damaged in this manner forms – often very brittle – hyperkeratoses, which are frequently the source of infections. These can occasionally lead to ulcers and osteomyelitis, which can even cause the loss of hands and feet. Changes such as these are occasionally found in diabetic polyneuropathies, but frequently in hereditary sensory neuropathies.

Anhydrosis symptoms frequently fail to be recognized. By contrast, normal sweating is often perceived as abnormal. Therefore, in any sweat secretion disorder, the abnormal side must first be found and then it should be decided whether damage is preganglionic and thus central, or peripheral and thus postganglionic. Since in a root lesion only the preganglionic segment in the root can be damaged, a sweat secretion disorder never occurs with motor and sensory deficits. However, all three functions are abnormal if the postganglionic segment is absent when the nerve is severed.

Sweat test

Disorders of the vegetative nervous system from peripheral nerve damage can be most easily determined by a sweat test. For this, not quantitative sweating, but rather qualitative sweating is tested. If the clinical examination shows marked sweating, the test can be dispensed with; the test is performed only in case of doubt or to document a finding.

Electrophysiology

Determining nerve conduction velocity (NCV) and an EMG examination are valuable techniques in addition to clinical examination findings. However, using them is meaningful only if a clear clinical question is formulated (Figure 22).

With the aid of an EMG examination, the following questions can be answered:

1. Is a nerve severed, or is there a conduction block?

   This question can be answered with certainty only 10–14 days after a nerve lesion in conjunction with an NCV examination. If with an EMG examination spontaneous electrical activity is found, this always indicates severance, while electrical silence indicates anatomical continuity with a functional electrical conduction block.

2. Are there signs of reinnervation?

   Before clinical signs of reinnervation with movement in the muscle, there are myographic reinnervation potentials. If there is voluntary innervation, there is a clear decrease in spontaneous activity and first small, low-amplitude, and then wide polyphasic and higher-amplitude potentials can be registered.

3. Are there any indications of neurogenic damage?

   Partial nerve lesions or root damage leads to only a partial innervation disorder of the muscle. In voluntary innervation, there are biphasic potentials that combine to a dense summation pattern in a healthy muscle. In partial denervation, the dense interference does not occur. In addition to spontaneous activity, reinnervation potentials can also be found depending on the duration of the damage. Such changes can also be found in polyneuropathies.

Nerve conduction velocity is helpful to decide the following (Figure 23):

1. Is conduction velocity impaired?

   Motor and sensory nerves can be tested for conductivity separately. For this, a summation action potential is recorded either directly from the nerve or from the attached muscle. The fiber segments that conduct most quickly form the start of the action potential, so that the NCV provides data only for the most rapidly conducting fibers. But the other action potential segments are also of considerable diagnos-

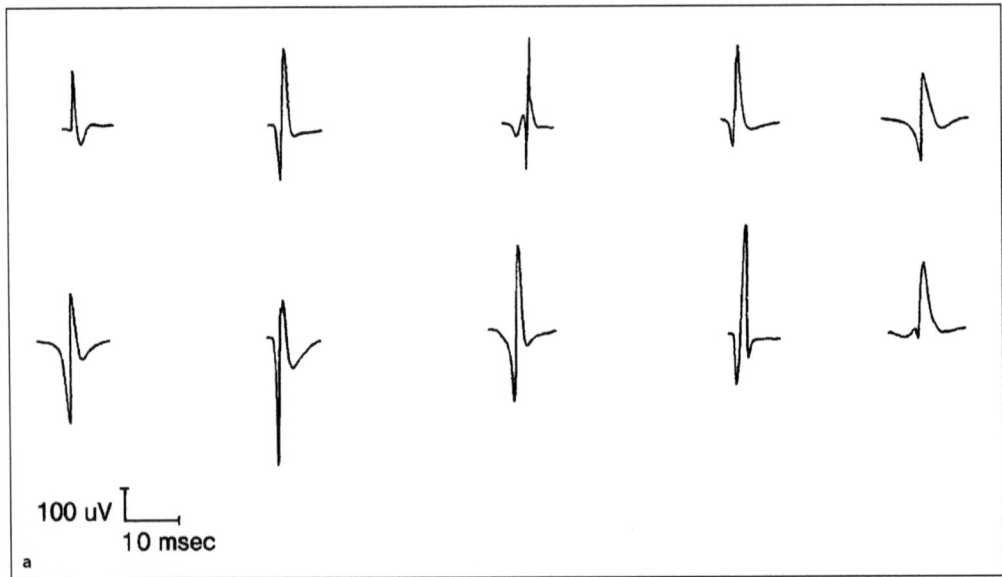

100 uV

10 msec

a

Figure 22 EMG potentials of normal and diseased muscle. Action potentials of motor units as they are shown in an EMG recording. In **a**: normal configuration of biphasic units. In **b**: polyphasic units, which in this frequency indicate neuromuscular damage. In **c**: typical features of denervation – positive sharp wave and biphasic fibrillation potential

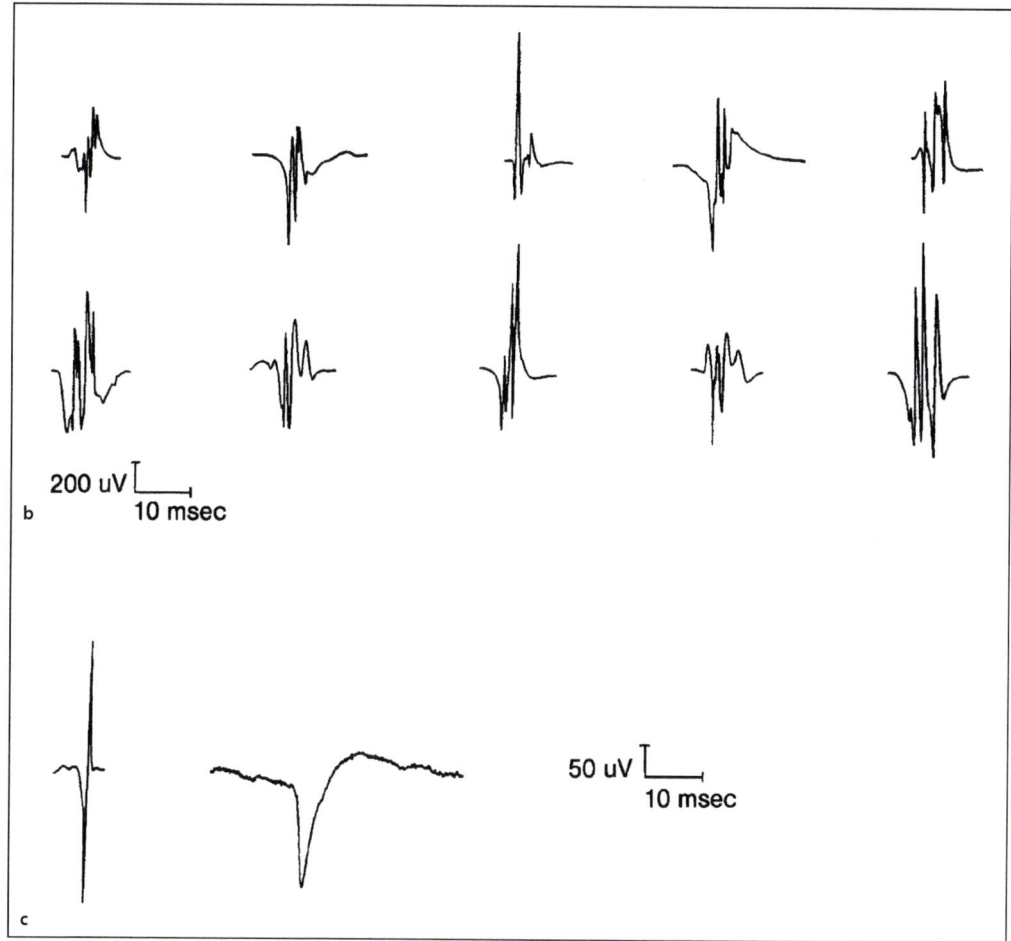

200 uV
10 msec

b

50 uV
10 msec

c

Figure 22b,c

tic importance. As the sum of action potentials, they consist of all fibers, from the fast to the slowly conducting fibers. Changes in the usually biphasic summation action potential with respect to the duration, form, and amplitude of the potential reflect the changes of conduction velocity of the nerve fibers over time and are very meaningful parameters in an NCV examination.

2. Is there a conduction block or circumscribed damage?
A conduction block is a circumscribed nerve lesion that affects only the myelin sheath (Figure 5) or electrolyte channels. If there is

electrical stimulation of the nerve above the lesion, no action potential, or only a considerably changed potential, can be registered distal to it. However, conduction is normal if the nerve is stimulated below the lesion. To record local changes, the nerve is often stimulated stepwise or fractionated. Stimulation of the upper arm, above and below the ulnar nerve groove, and at the wrist allows a lesion along the ulnar nerve to be found.

If no changes are found in the EMG or NCV examinations, other causes of the lesion must be considered, such as those listed in Table 2.

3

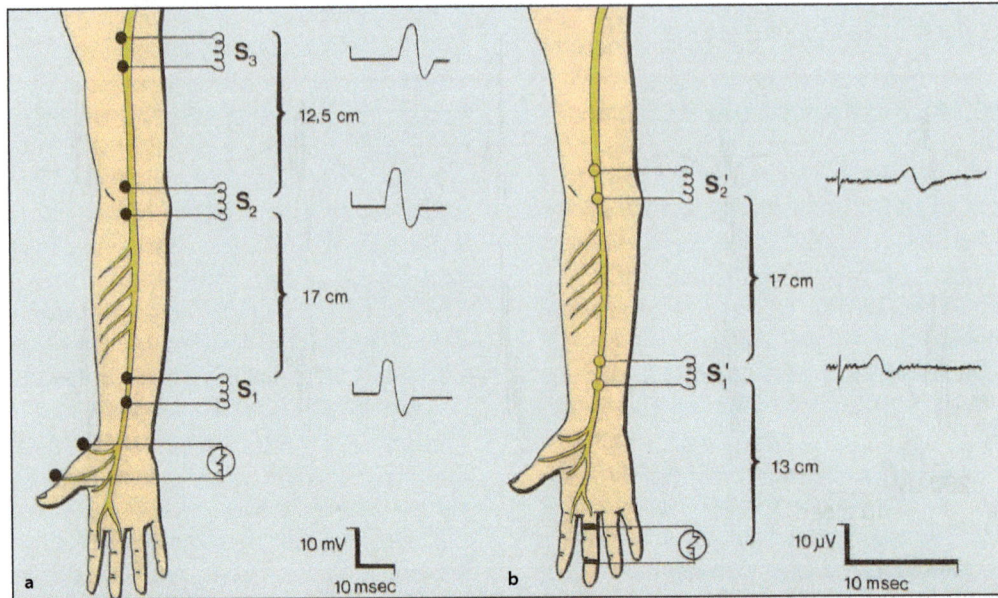

Figure 23 Technique for studying motor or sensory conduction velocity of the median nerve. Electrical stimulation of the median nerve is performed in **a** in S1–S3. The orthodromic impulse traveling down the nerve generates a muscle action potential in the thenar. Recording electrodes attached to the thenar pick up the compound action potential of the muscle. The effect of stimulation is depicted next to the stimulating points S1–S3. As the distance between the stimulus and the ball of the thumb decreases, motor latency is shortened. Motor latency also includes processes of neuromuscular transmission and the spread of excitation in the muscle. To exclude these factors, when calculating the NCV, the difference of two motor latencies (S2–S1) is calculated reflecting the conduction time between the stimulus points S2 and S1. The distance (17 cm) divided by the conduction time yields the NCV (forearm). The form and amplitude constant of all three muscle response potentials is an important criterion. In **b** the nerve is stimulated at S'1 and S'2 (antidromic stimulus = against the nerve conduction direction) and its action potential is measured on the middle finger. The latency between the stimulus and beginning of the nerve action potential is the conduction time. Distance : time = NCV

Radicular lesions

4

4.1   Segmental innervation

Radicular lesions have a deficit pattern that differs from that of peripheral nerve damage. This difference is presented using the example of sensory skin supply in Figure 13. In analogy to this, motor supply of the musculature also differs. For example, when the axillary nerve is severed, complete paresis of the deltoid muscle occurs, but severance of the C5 root leaves some remaining innervation in the clearly paralyzed muscle. This incomplete paresis is the result of the radicular supply principle, in which one muscle is never completely supplied by one segment, but rather several muscles are partially supplied by it. The muscle that receives the most nerve fiber inflow shows the greatest paresis from a radicular lesion and is therefore designated the indicator muscle of the segment. Indicator muscles thus are a guideline for recognizing the level of radicular damage.

Lateral disc herniations tend to affect the dorsal root, so sensory deficits and pain are the main symptoms. If only the ventral root is damaged, the innervated muscles, in particular the indicator muscle, are affected. There are always reflex disorders, since in both cases, either the supplying sensory arm or the efferent motor segment of the reflex arc is damaged. Symmetrical innervation of the body allows an assessment to be made by comparing the reflexes of the two sides, so that the evaluation of a weakened or absent reflex is easier to make. In contrast to peripheral nerve lesions, no sweat secretion disorders are expected in radicular injuries. The postganglionic fibers proceeding from the sympathicus trunk receive so much synaptic activation in the sympathicus ganglion from preganglionic fibers located above and below the segment that a monoradicular preganglionic deficit is completely compensated.

Radicular lesions generally arise either acutely from a herniated disc or gradually from degeneration due to osseous projections of the spine and neuroforaminal narrowing. In both cases, the pressure usually damages the nerve fibers in the roots. However, irritation can also be caused by poor circulation in the radicular artery supply territory. The narrowing of the lumen of the artery due to pressure can lead to ischemia followed by edema. This constant mechanical and metabolic irritation triggers proliferation of inter- and intrafascicular as well as endoneural connective tissue and causes strictures. Another radicular lesion can be triggered by mechanical and inflammatory irritation, which initially affects only the pia mater near the root and can lead to treatment-resistant arachnopathy through adhesions. These changes also affect the sinuvertebral nerve, which innervates the nerve root sheath and is involved in radicular pain via nociceptors. We know now that there are also inflammatory processes that attack the nerve fibers and the dorsal root ganglion (DRG). The inflammatory content of the nucleus pulposus plays a major role. It is considered a very important pathophysiological factor in the mechanisms of chronic sciatica pain. It has been demonstrated that the application of nucleus pulposus material from a herniated disc can induce pronounced morphological and functional changes in the adjacent neurological tissue. One of the pro-inflammatory cytokines that is a principal modulator has been identified as TNF-α. Inflammatory mediators such as prostaglandin E2, serotonin, adenosine, bradykinin, and epinephrine produce hyperalgesia through activation of protein kinase A and protein kinase C in primary afferent neurons.

TNF-α has recently been shown to be a constituent of nucleus pulposus in vivo and herniated nucleus pulposus in culture. This cytokine is one of the pro-inflammatory substances and a principal modulator of the early degenerative changes during peripheral nerve injury. It is therefore a key pathogenic factor in the initiation and maintenance of neuropathic pain states. In animal experiments it produces thermal hyperalgesia and mechanical allodynia that are qualitatively similar to the pain behavior observed in nerve injuries with Wallerian degeneration. The inflammation or nerve injury can trigger a sequence of events that lead to early, intermediate, and late changes in gene expression that can cause the development of acute and chronic pain. The alteration in gene expression and the resultant changes in the excitability of DRG neurons may be associated with peripheral and central sensitization in acute and chronic pain states (see Sect. 6.2).

The anatomy of segmental innervation (Figure 24) explains the damage pattern that occurs in a

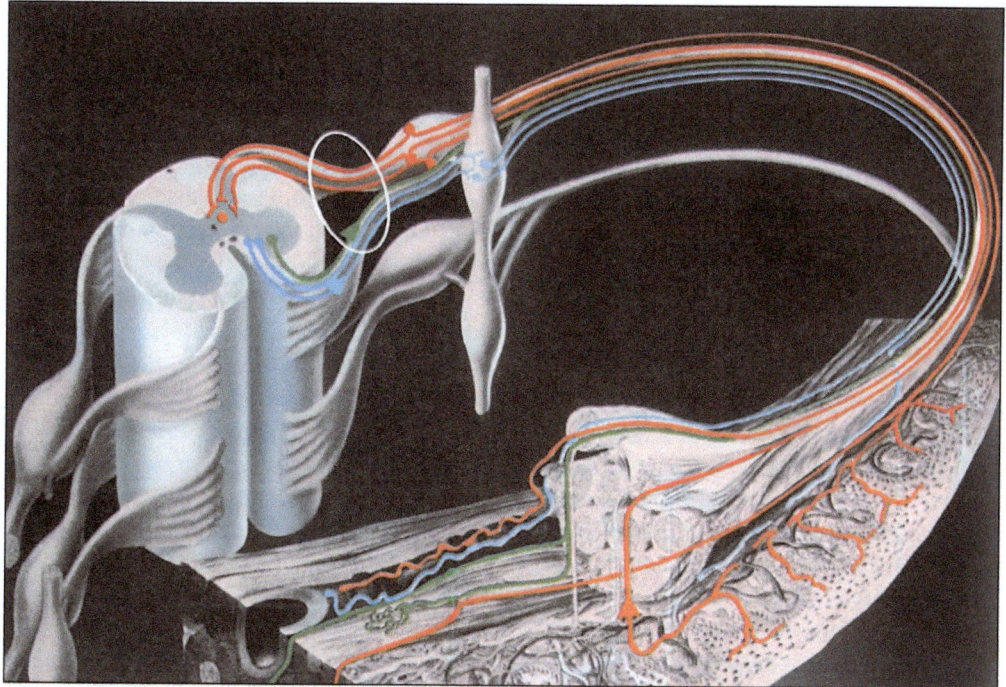

Figure 24 Schematic drawing of segmental innervation. The white oval contains the structures that can be acutely damaged in a lateral or intraforaminal disc prolapse or chronically damaged by degenerative changes to the vertebrae. Damage to the dorsal root can either interrupt the afferent sensory fibers (red) – with a corresponding loss of sensitivity in the dermatome – or irritate the fibers, causing pain. The motor pathways (green) are in the ventral root and their disorders lead to muscle paralysis. Damage to the sympathetic preganglionic fibers (blue), which are also located in the ventral root, has no consequences in the dermatome, as other segments compensate the postganglionic neuron in the sympathicus trunk

radicular lesion. When the motor axon is severed, Wallerian degeneration begins in the nerve up to the muscle. If the nerve is to innervate this muscle, it must regenerate over the entire degeneration area. These distances are often shorter in sensory nerves, as the nerve is usually damaged only in the segment that leads from the spinal ganglion to the dorsal horn. Thus only the proximal neurite is damaged and the peripheral axon remains intact. However, when centrally directed axon segments regenerate, their myelin sheath and nerve conduction characteristics are also changed. This is significant for the development of chronic pain and is addressed in detail in Chap. 6, "Acute and chronic pain." It is important to know here that during movement of the spine, the roots are always displaced. This alone can cause the fibers that are now sensitive to mechanical irritation to generate impulses; their activity is inter-

preted as pain by the central system. Moreover, even the smallest lesions in the dorsal root can lead to intraneural scarring, strictures, and neuromas. The attempt to intervene in the dorsal root structures by microsurgery is highly problematic, as every procedure makes new lesions that can be the start of more scars and further degeneration.

The clinical features of radicular damage are:

- Motor paresis of the segmental indicator musculature
- Sharply circumscribed radicular hypoalgesia
- A wide hypoesthetic zone extending beyond the dermatome
- Reflex disorders
- Absence of sweat secretion disorders

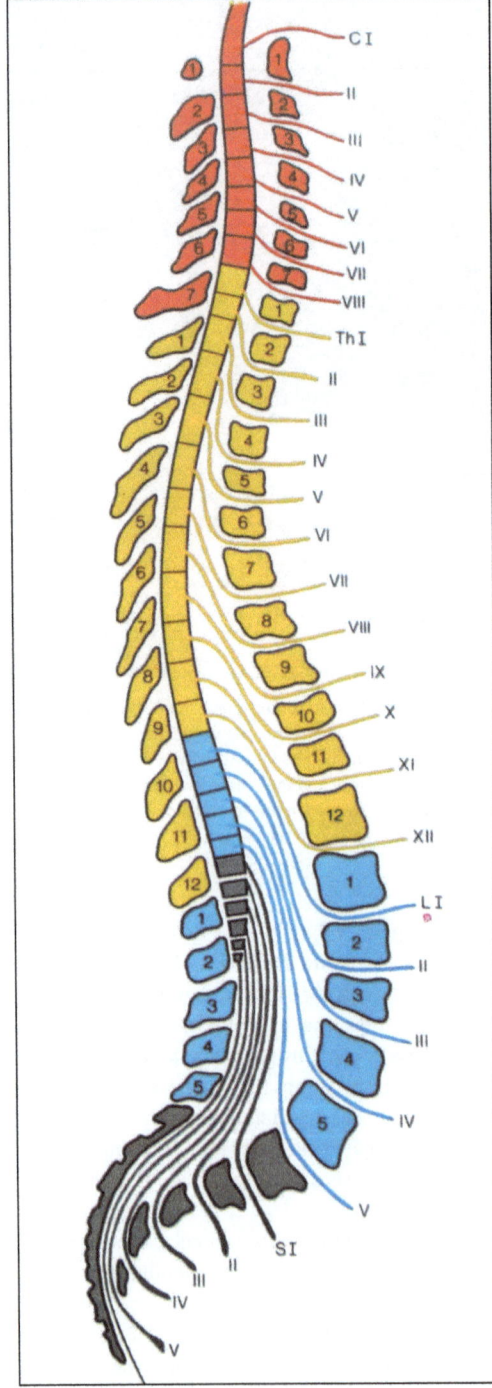

Figure 25 Topographic relation of spinal segments and nerve roots

4.2   Remarks on terminology

The description of root damage sometimes leads to problems with nomenclature. As a general rule, the damaged nerve root is always related to the spinal segment from which it proceeds. The allocation can be made easily in the diagram in Figure 25.

To understand the degenerative changes, their radiological correlates, and the possibilities for surgical intervention, it is useful to recapitulate a few details. The vertebra, its processes, the small vertebral joints and their facets, the disc, and the ligaments form a unit, the functional movement segment (Figure 26). Changes in it are not isolated, but affect in particular the adjacent and other segments as well. Problems can arise if the connective tissue does not stabilize the vertebral segments properly, resulting in hypermobility. This can cause pain. Overuse or malposition of the vertebral joints, which can irritate or damage the joint facets, can also trigger a painful facet syndrome.

Pivot and stress points on a vertebra and joints are often the starting points for degenerative changes that can be imaged by radiology. In the terminology of these changes, chondrosis is understood to be the calcification of the vertebral end plate (osteochondrosis). If the ligaments exert additional pressure on the vertebrae, this can lead to the formation of a bony spur, spondylosis. If the vertebral joints on the joint facet are sclerotic, there is arthrosis; if the surfaces also display apophyseal degeneration from bony spurs, we speak of spondyloarthrosis. This can lead to narrowing of the intervertebral foramina and compression of the spinal nerves, particulary in the cervical spine. With this degeneration, only slight disc protrusion can increase the narrowing and trigger pronounced radicular symptoms. The same applies to the spinal canal leading to compression of the spinal cord. Herniated discs are described as hard if they involve considerable degenerative changes of the bone. By contrast, cartilaginous sequester is described as a soft herniation.

Degenerative changes to the disc also have firmly defined designations. Degenerated, but still intact annulus fibrosus protruding into the spinal canal through the nucleus pulposus is a protrusion. Only when the annulus fibrosus is torn and the nucleus pulposus exudes from the intervertebral space do

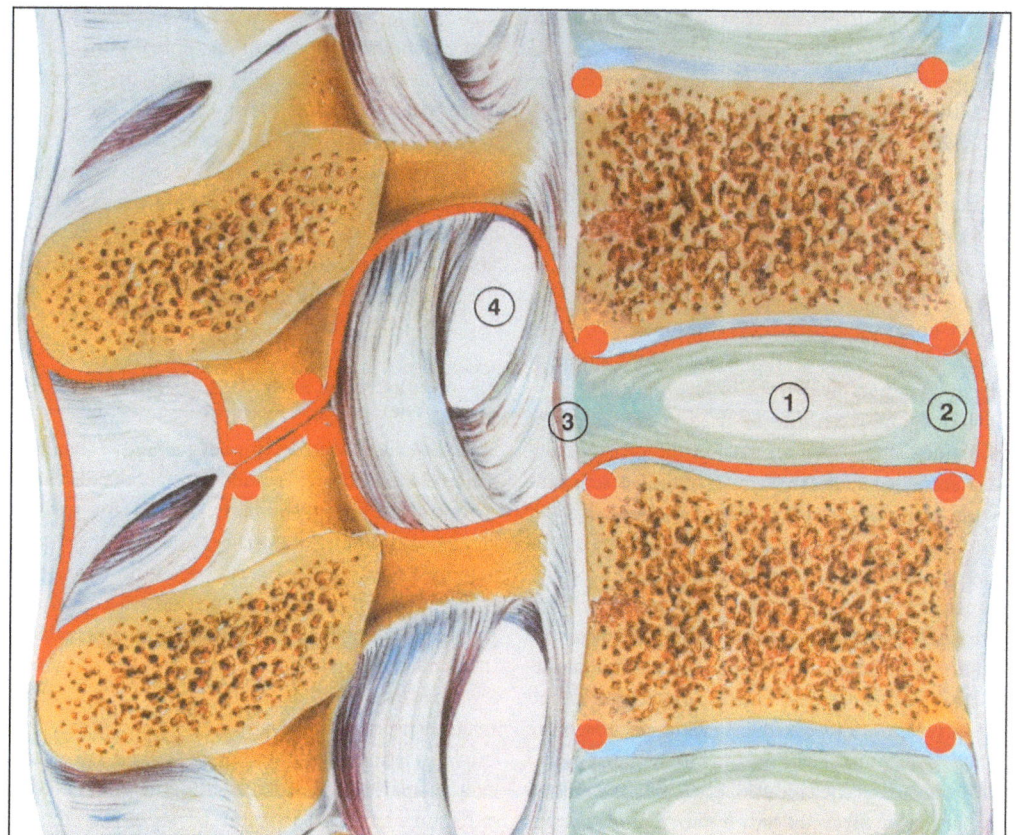

Figure 26 The functional unit of movement. The red-bordered regions form a unit, the functional unit of movement. Red dot = Pivot and stress points on the vertebra. 1 Nucleus pulposus, 2 Annulus fibrosus, 3 Longitudinal ligament, 4 Intervertebral foramen

we speak of prolapse. Clinically, the two lesions produce the same symptoms. The differentiation of disc protrusions into subligamentary, perforated, or sequestered is not significant for the clinical symptoms, but it is relevant for making decisions on treatment.

To understand the difference between acute and chronic back pain it is important to know the changing innervation pattern of the disc in these conditions. Afferent innervation of the intervertebral disc takes place in healthy subjects only in the outer third of the annulus fibrosus. Nociceptive A-δ and C fibers are involved. Similar innervation has been identified in the anterior and posterior longitudinal band, which also contains proprioceptive fibers of pacinian corpuscles (Figure 18) and other mech-

anoreceptors (Figure 17). In degenerative disc disease and low back pain, the situation is very different. Within this pathology the nociceptive fibers from the longitudinal bands and annulus fibrosus start to sprout and extend into the nucleus pulposus, extending deeply into the intervertebral disc. Nerve fibers protruding into the degenerated intervertebral disc along with blood vessels have also been identified. In chronically degenerated discs, the compression and irritation of free nerve endings (Figure 16) will give rise to nociceptive pain. Inflammatory changes within the disc or due to compression of the nerve root by the disc or degenerative change of the bony structure induce neuropathic pain. The parallel existence of the two types of pain forms the clinical syndrome of sciatica. The mixed pain concept mirrors this coexistence and interaction of the two pain forms and helps to understand the pharmacological therapeutic approach (see Sect. 10.3.8).

If there is sufficient suspicion of a discogenic cause of pain without verification of a corresponding morphological correlate in non-invasive diagnostics, discography [perhaps with post-discography computed tomography (CT)] may be useful. In this method, the increased pressure in the disc can induce nociceptive pain. This is sometimes used to verify an intact annulus fibrosus before applying percutaneous surgical procedures.

Discography can thus be used in case of doubt to distinguish between protrusion and prolapse. On a CT scan, degenerative changes of the nucleus pulposus indicate entrapped air or could also result from extruded disc material. The presence of this linear radiolucency in the disc space (vacuum phenomenon) is a typical finding of degenerative disc disease. Depending on localization, the prolapse can be medial toward the lumen of the canal or lateral toward the foramen. The disc material extruded in a prolapse is called sequester and can reach the spinal canal, remain in the segment, or glide upward or downward in the epidural space. Occasionally, a protrusion involves not only a radicular lesion, but perforation of the dura as well and the sequester enters the subarachnoid space, possibly sinking sacraly, where it dissipates. Such sequester slippage can thus lead to lesions in several segments, although the corresponding disc is still intact. Sequester rarely slips so far that it cannot be seen in radiology diagnostics.

The cervical spine consists of seven vertebrae, but eight roots exit from the cervical cord (Figure 25). In the cervical region, the respective roots exit above the vertebra so that a herniated disc in segment C5/C6 affects root C6. However, in the lumbar spine, the roots exit below the vertebra, so that a herniation at L4/L5 damages root L4. As the spinal cord is almost a third shorter than the spinal column, the vertebrae and spinal cord segments are offset against each other. This must be taken into consideration when localizing the level of lesions and explains why a medial prolapse in the lumbar spine can affect the conus, a part of the spinal cord, only at the level of the 1st lumbar vertebra. Below

that are the bundle-like structures of nerve roots that extend caudally from the spinal cord. They form the cauda equina, followed by the terminal filum which is attached to the sacrum. These anatomical features make it impossible to reach the spinal cord in a lumbar puncture, as many patients fear. Nerve roots can be touched during the puncture, but they immediately evade the needle, as they are floating in fluid, not fixated. However, contact can trigger shock-like radicular pain.

4.3   Overuse and disc herniation

Since disc disease arises from degenerative changes, its development is never sudden. Only the disc herniation itself occurs suddenly. Its pain and paresis can be so dramatic that the patient often views the activity he was just performing as the triggering event. Thus, especially for people who perform physically demanding labor, the prolapse, especially if it occurred at work, frequently leads to protracted legal battles, which generally can prove no causal relationship between work and the disease. If the causal relationships between cause and effect are ever properly linked, it usually emerges that the situation associated with the prolapse is a coincidental finale of a slowly progressive degenerative process, but not the cause of the process. How varied the moment causing the prolapse can be is proven by the fact that sometimes patients are surprised by its occurrence during sleep or when they wake up in the morning. However, the cause of a prolapse must be assessed entirely differently if bone or ligament damage from a violent trauma suddenly occurs that leads to a displacement of the disc (Figure 27).

4.4   Diagnostics

The patient history and examination give a first impression in the event of an acute occurrence. Leading symptoms are neurological deficits such as changes in reflexes, muscle paresis, and sensory disorders. It is not quite so easy to answer the question of an additional acute occurrence for patients with chronic radicular pain. This can often be clarified only clinically, as the radiology finding is useful for diagnostic purposes only in conjunction with the clinical impression.

To objectify the clinical findings, EMG-examinations are often ordered. In denervation, this test can identify the muscles involved very well and help find out whether their deficit pattern is radicular or caused by another disorder. However, an EMG does not help in the acute stage of a herniated disc. Since the nerve is initially degenerated after the loss of function, but the stump still maintains trophic contact with the muscle, signs of denervation do not occur until 10–14 days later. The situation is also complicated if a recurrence is suspected after disc surgery. Changes in the EMG do not show whether the damage is acute or chronic. This diagnostic dilemma also occurs for patients who suffer from both polyneuropathy and herniated disc. Both illnesses can lead to denervation of the muscle or to enlargement of its motor units. There is no specific muscle reaction for the respective disease. Here also, findings from technical examinations can be properly assessed only with the clinical impression.

There are many reasons for unsuccessful surgical or conservative therapy that must be taken into consideration, especially before invasive treatment, and surgery in particular. For unclear pain such as a chronic cervical syndrome or lumbago, diagnostic radiology is carried out. Depending on the age of a population, CT or magnetic resonance imaging (MRI) examinations can reveal asymptomatic disc herniations in every second to fourth patient. If a patient complains of pain, a correlation between the patient's description and the radiological finding is quickly made.

Only when the clinical finding and the technical examination fit can therapeutic action ensue.

In radiological diagnostics, the value of native diagnostics has decreased with the increasing improvements in the quality of imaging techniques such as CT and MRI. In the native image, soft tissues can be imaged poorly, in addition, distortion or overlays of anatomical structures due to projection occur. It can still be useful to rule out fractures, to quantify a potential spondylolisthesis, or to clarify the number of lumbar vertebrae.

Prior to every x-ray, it is necessary to specify a level of radicular lesion. Only in this manner can a targeted CT or MRI of the individual segments be carried out and the radiological findings together with clinical findings yield a diagnosis. In case of clearly monoradicular symptoms, a CT of the lumbar spine of the affected and two adjacent segments is sufficient. The CT examination provides an especially good image of the osseous structures. Information about the bones is important if tumors in the form of metastases are being considered as a differential diagnosis. The disadvantage of CT in comparison with MRI is exposure to radiation (2-mm slices are required).

If the localization is imprecise, a large section of the spine must be examined. MRI makes it possible to image longer segments of the spinal canal with bones and tissue. If the level of the lesion is unclear, MRI or spiral CT is first use; it has replaced myelography to a great extent.

Myelography of the spinal canal is still used if an MRI examination is not possible. This can be the case if the patient has a pacemaker or metal implants. Occasionally, an MRI/CT scan can be insufficient for a diagnosis if there are complicated conditions. For verification of a relevant functional narrowing and documentation of whether there is still cerebrospinal fluid flow, a lumbar puncture must be made and contrast medium injected into the subarachnoid space. This issue can arise if there is stenosis of the spinal canal with narrowing at several segments or if there is pronounced scoliosis. Further subtle diagnostics can be carried out with CT myelography. The post-myelography CT aids in accurately imaging compression of the dural sac and nerve roots exiting there.

An imprecise request such as "CT of the lumbar spine" is problematic. The radiologist, not knowing the clinical situation, records the minute details of every degenerative change. The colleague who requested the CT gratefully accepts every description to confirm his general suspicion of disc damage. The patient suffers when his lumbago is interpreted as a herniated disc because of slight protrusions in the CT. Since CT and MRI are often used liberally in diagnosing lumbago, such misinterpretations are not infrequent.

4

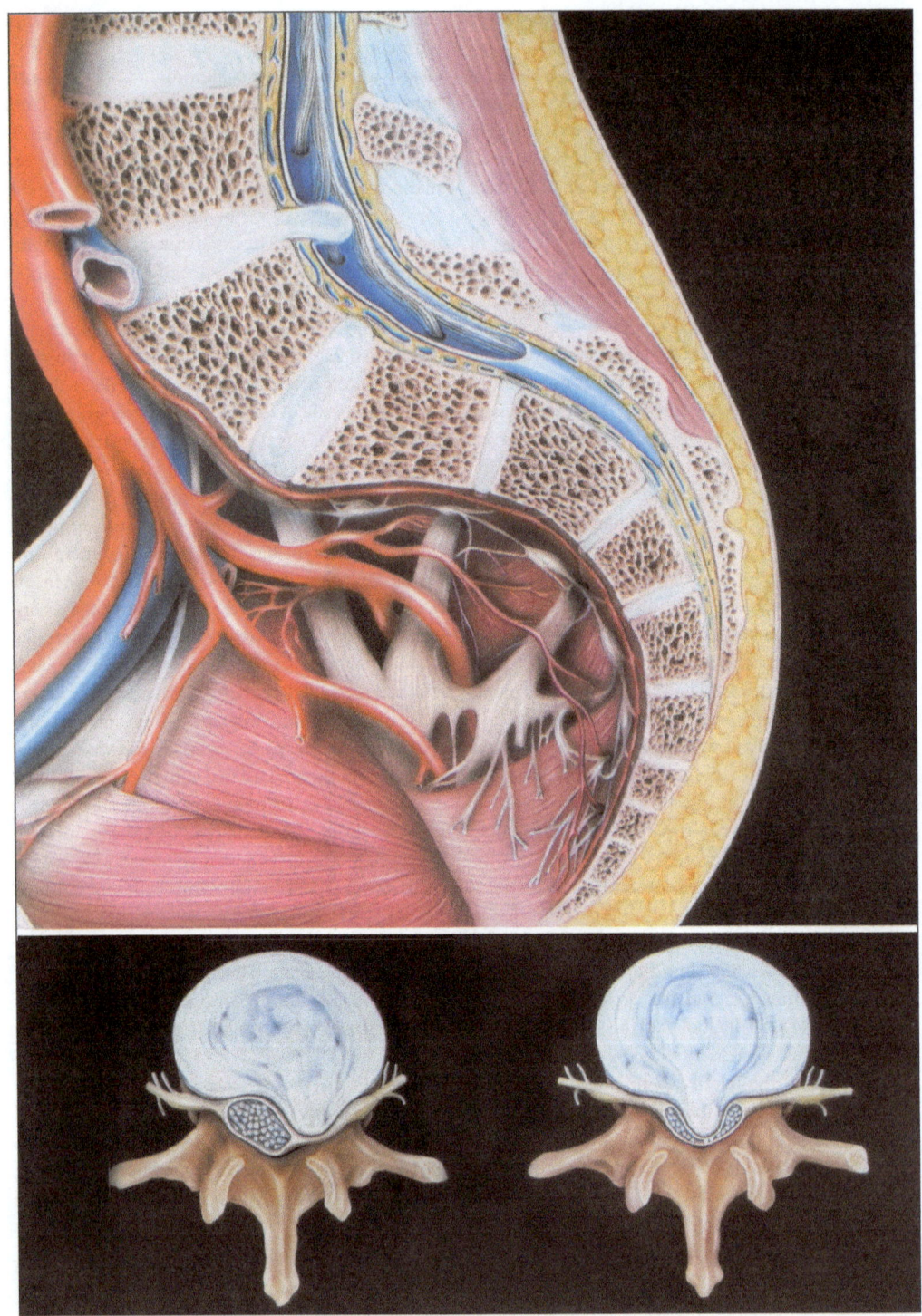

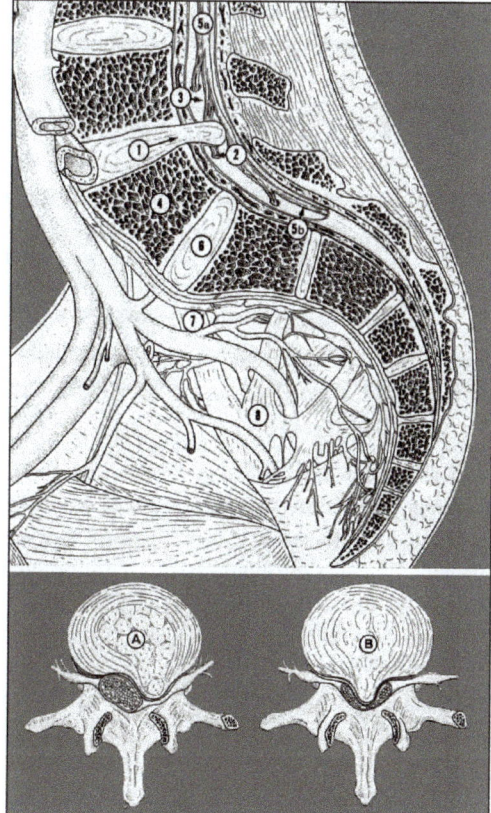

Figure 27 Anatomy and mechanics of disc herniation. B medial disc prolapse, 2 4th lumbar root, 3 subarachnoid space, 4 5th lumbar vertebra, 9 epidural space

4.5   Therapy of a herniated disc

Clinical experience has shown that up to 90% of symptomatic herniated discs can be controlled by conservative measures. However, an acute herniated disc with radicular paresis usually, but not always, needs surgical intervention. An operation leads to quicker recovery from the impairment due to pain than conservative treatment, but the long-term results are similar. It is still unclear today what effect an operation has on the course of natural degenerative changes in the spine.

An acute prolapse always requires rapid clarification. Medication is useful only after it is clear what triggered the pain. Any awkward positions assumed by patients are simultaneously protective postures

that offer a certain degree of pain relief. Medication should be used only to relieve pain; under no circumstances should pain relief delay diagnostics or even worse, obscure the situation. For as soon as a root loses its function from the pressure of sequester, the root dies and this is associated with a misleading improvement in clinical symptoms.

When the causes of the symptoms have been found, the pain can be treated with medication until the operation or start of conservative treatment. Cortisone injections of about 80 mg dexamethasone on day 1 and tapering off over the next 5–7 days have been shown to be helpful. Cortisone decompresses the root with an anti-edematous effect and usually gives relief within a few hours; in addition, antiphlogistic analgesics such as acetylsalicylic acid can be given.

The necessity of surgical intervention must be determined individually for every case. For cervical disc herniation, surgery should be considered if treatment-resistant pain persists over 6–8 weeks despite intensive conservative measures or if neurological deficits are progressive.

There is an indication for immediate surgery for a caudal syndrome; these symptoms are always to be treated as an emergency. Cases with clear and increasingly severe paralysis should be assessed the same way. A strength level of 3 Medical Research Council (MRC) (Table 3) is considered severe paralysis. The indication for prompt surgery can also be given if there is massive radicular pain with a corresponding finding in imaging that cannot be brought under control with analgesics.

There are many different operative techniques for treatment. However, anterolateral access to the spinal column is usually chosen for removal of a cervical disc herniation. In this method, access to the vertebra is made via a lateral incision at the neck. The respective vertebral space is located and a drill is then set to allow the disc space and segments of the adjacent vertebrae to be opened. Subsequently, disc residue and uneven osteochondrotic margins can be removed through this drill hole. Finally, a bone plug is taken from the iliac crest and pressed into the expanded drill hole. This intercorporal spondylodesis quickly stabilizes the spinal column and the intervertebral foramina remain spread slightly apart. The bone plug links the two vertebrae

4

by ossification; postoperatively, a Schanz cervical collar must be worn for a few weeks during this process. More rapid stabilization can be achieved with titanium or plastic plugs. But for young patients the side effect of this technique should be considered, namely, mechanical overuse of the adjacent discs. Observations over a decade showed that in 25% of patients, another operation for a reoccurrence of disc herniation became necessary.

The lumbar prolapse is operated on from the back. Today, experienced surgeons choose microsurgery access using an operating microscope. This access causes fewer traumas to surrounding muscle tissue than a wide exposure of the spine and shortens recovery time considerably. Using a smaller incision, the back musculature must be displaced only far enough to allow access for the microinstruments. After removal of the ligamentum flavum, and if needed, parts of the vertebral arches (interlaminary fenestration) the spinal canal is open. The dura mater is held to one side, the intervertebral space is located, and the annulus fibrosus is incised. The nucleus and the degenerated annulus fibrosus are removed using bone curettes. Any displaced sequester must be located and also removed. The intervertebral space remains empty and will soon be filled with connective tissue. Temporary urinary dysfunction may occasionally occur due to an intraoperative irritation of the vegetative fibers.

For the lumbar region, an alternative treatment option is chemonucleolysis. Under x-ray guidance, a proteolytic enzyme is injected into the intervertebral space. This collagenase dissolves the disc so that it can be resorbed. The processes needed for proteolysis require water, which is initially stored in the disc, making it swell up, leading to an increase in volume for 1–2 days. In addition to considerable pain, the increase in volume can also aggravate already existing radicular symptoms. This risk cannot be taken if there are severe neurological deficits; the method is thus contraindicated for such cases. In addition, for this method, the annulus fibrosus of the disc must be intact so that the collagenase cannot leak out of the disc, as contact with the spinal cord would damage it. The procedure is not always successful, sometimes making it necessary to operate later anyway. Since collagenase is contained in many common household meat tenderizers, this

contact may have already led to sensitization of the patient and the therapeutic injection can trigger an allergic reaction. The method is thus not entirely unproblematic and is therefore hardly used now. A comparison of open discectomy with chemonucleolysis shows that the surgical option yields better results. Therefore chemonucleolysis is no longer a treatment option.

Percutaneous nucleotomy is another option. It can be performed on an outpatient basis and is thus frequently propagated. In this procedure, the damaged disc is punctured with a rather thick hollow needle under x-ray guidance and disc tissue is removed endoscopically. This method has thus far been compared only with chemonucleolysis; chemonucleolysis proved to be more advantageous.

Under the impression that a disc prosthesis in the lumbar region improves the function of the spinal column, such prostheses have been implanted for many years. The indication for this operation is often chronic back pain and verified disc degeneration. Although there are already many years of experience with the prostheses, scientific studies have not yet proven the effectiveness of this procedure.

4.5.1 Conservative or operative therapy

The less the extent of neurological deficits can be objectified, the less urgent an operation becomes. But if the deficits are large, surgery is urgent. Urgency also decreases quickly if several days have passed since the lesion, even if the symptoms are severe. At this time, the damage is usually irreversible, acute relief to save the root is already too late. It remains a subjective, temperament-related clinical decision whether operating or pursuing conservative treatment is the better option. For chronic radicular disorders, in particular if pain is the main symptom, this question is especially problematic. Long-term pain shows again and again that the organic problems that trigger it become less important and increasingly, problems of chronification determine the dynamics of the illness. In addition, chronic lesions have often already caused such extended interfascicular and intrafascicular damage to the root that surgery to relieve pain does not have any effect. For these reasons, it is certainly wrong to assume that every herniated disc requires surgery.

If surgery is performed in the acute phase of disc herniation it can be expected that postoperatively there will be improvement in about 30% of the reflex changes, 70% of the sensory and motor paresis, and 80% of paresthesia. Quick improvement is seen for about 90% of reactive changes such as the cervical syndrome or lumbago. Not infrequently, the preoperative pain remains in a mild form, especially if there have already been chronic radicular lesions in the past. About 10% of all patients must undergo relapse surgery. Such relapses need not be the result of a poor initial operation. They are the result of further degenerative changes to the annulus fibrosus, which could not be completely removed in the initial operation if it was still largely intact.

Conservative treatment is done in stages. Initially, bed rest must be maintained until pain is reduced under steroid and non-steroid antiphlogistic therapy. Only then is the patient amenable to physical therapy. During this therapy, the patient should learn exercises to strengthen and relax musculature in order to continue them alone later. In addition, movement patterns should be trained to help reduce constitutional poor posture or avoid habitual movement errors. It is essential to first inform the patient about his illness and point out that active measures to strengthen the muscles can improve the stability of the back and improve movement patterns. Information from the physician is one of the most important components of prevention and therefore of the patient's coping with the situation.

4.5.2 The "failed back syndrome"

The failed back syndrome is a poorly defined term for chronic pain, which is also known by terms such as post-nucleotomy syndrome or post-discectomy syndrome.

This syndrome does not describe a patient population with homogeneous symptoms, but they usually share the following problems:

1. Postoperatively, the pain is unchanged at the preoperative level with or without a painfree interval.

2. The patients feel that the operation was unsuccessful.
3. With respect to functional capacity, in particular ability to work, postoperative deterioration is perceived.
4. Preoperative psychological tests give indications of inadequate ability to cope with illness and pain.

When studying the cause of the problem, one consideration should be that an asymptomatic prolapse was operated on under a false indication. It can also have been at the wrong level, the compression could be caused by an unrecognized osseous degeneration.

Or it was only a pseudoradicular problem. The cause of the pain is not a herniated disc, but primary or surgery-related segment instability, a facet syndrome, epidural fibrosis, or arachnopathy. In the operation, free sequester can have been overlooked, the osseous decompression was insufficient, or dural and nerve root damage occurred. Also, in the organic indication for the operation, which may have been lacking or in sufficient, the problem of chronification may have been overlooked and the operation thus became another component of iatrogenic chronification.

If there is no clear indication for surgery, conservative therapy is first recommended, which should, however, yield pronounced improvement of pain and increase in function within a period of 6–8 weeks. Otherwise, a change in the treatment plan or surgery should be discussed. The uncritical continuation of conservative measures entails the risk of chronification of pain.

Studies of the causes of chronification of back pain have shown that psychological and psychosocial aspects were more significant parameters than the organic findings. It was also shown that the psychological and psychosocial parameters could be used as predictors with 80% prediction probability for the course of the illness and rehabilitation results. The prevalence of psychological and psychosomatic disorders was 20–25%, so that for patients with chronic back pain, psychological comorbidity must always be taken into account in the diagnostic and therapeutic concept. Psychological comorbidity will always have an influence on the course of the illness. Monocausal and purely somatic treatment strategies have only limited success in the treatment of chronic pain. All of this suggests that to prevent chronification, organic and psychosomatic diagnostics must form the basis of a treatment plan. Negative factors of influence are diagnoses such as anxiety, depression, and a somatoform or personality disorder. Ongoing conflict situations at the workplace and in private life, heavy physical labor, and legal proceedings or the expectation of compensation were also found to complicate the situation.

4.6   Cervical root syndromes

For neck and shoulder pain, the patient history and examination should first attempt to determine whether there is an isolated cervical root irritation and whether it is a multisegmental process or a pseudoradicular process. Generally, in the cervical root syndrome, the pain radiates into the arm, in cervicocephalgia it radiates into the arm and back of the head. In both cases, certain movements and positions of the cervical spine can provoke, trigger, or worsen the pain. This leads to characteristic mal-

posture with the shoulders drawn up and the neck held stiffly. The musculature in the shoulder-neck region is hard with palpable myogeloses. At night, when muscle tone is reduced and the antalgic posture from voluntary innervation slackens, pain increases (Table 5).

Increasing age leads to degenerative changes to the spinal column, which, however, need not become symptomatic. But if pain occurs, freedom from pain can be achieved for only about 35% of patients. If the patients are still working, recurring cervico-brachialgia can be expected, which despite extensive treatment results in some 20% of patients being unable to continue work.

The cervical spine must ensure movement of the head so that it retains the freedom of movement required for its sensorimotor tasks. Depending on individual predisposition, degenerative changes occur over the years in some people that are most pronounced in the segments of the spine that are subject to the greatest mechanical strain. This is the reason why most radicular lesions occur at the level of the most active segments C5/6 and C6/7. Roots C6 and C7 are most frequently damaged, and somewhat less frequently roots C5 and C8 (Figures 25, 30, 31). In all cases, root compression leads to pain ra-

Table 5 Summary of cervical root syndromes

Segment	Sensitivity	Indicator muscle	Reflexes	Differential Diagnosis
C3 – 4	Pain Deficits in the neck and shoulder region corresponding to segmental innervation	Paresis Diaphragm paresis (x-ray fluoroscopy)	Disorders not clinically apparent	
C5	Disorders of the shoulder, such as deltoid muscle region and upper arm	More pronounced in the deltoid muscle than in the brachial biceps	Biceps reflex reduced	
C6	Radial upper arm and forearm up to thumb	Especially brachial biceps muscle, less in the brachio-radial muscle	Biceps reflex reduced to absent	Carpal tunnel syndrome
C7	Inside of forearm and 2nd to 4th fingers	Especially triceps muscle, also hand and finger extensors, pectoralis major	Triceps reflex reduced to absent	
C8	Ulnar forearm and side of hand with the little finger	Similar to C7 but stronger paresis in the small musculature of the hand	Triceps reflex	Lesion of lower plexus, lesion of ulnar nerve

diating into the arm up to the hand that can be increased by movement of the cervical spine and relieved by wearing a cervical collar. If the anatomical situation changes and the nerve root is constantly compressed, the patient complains of unpleasant sensorimotor dysesthesia. In addition to reflex deficits, there are sensitivity disorders in the typical segment and finally numbness in the dermatome and paralysis of the musculature.

4.6.1 Cervical myelopathy

Herniated discs in the cervical and thoracic spinal canal act as space-occupying masses that can affect the root, root and spinal cord, or spinal cord alone (Figures 28, 29). Every cervical disc herniation, acute or chronic, can thus cause a transverse spinal lesion with spinal cord compression that must be quickly operated on. Only in a thorough examination of the patient can the initial symptoms of a cervical myelopathy be found – in case of doubt, a colleague's advice should be sought.

Degenerative changes of the cervical vertebrae frequently cause stenosis of the spinal canal. Cervical spondylotic myelopathy is the most common cause of damage to the cervical spinal cord in old age. Degenerative changes of the cervical spine are found in 75% of all people over age 65; however, only a small percentage of them develop myelopathy. Lesions of the spinal cord are encouraged by a constitutional narrowing of the cervical canal. Normal values of the spinal canal diameters are given in Figure 31.

Depending on where they are located, degenerative changes cause variable combinations of symptoms with radicular deficits of the upper extremities (spondylotic radiculopathy) and symptoms of spinal cord damage of the lower extremities (cervical spondylotic myelopathy) or a combination of the two (radiculomyelopathy). Every other slowly progressive space-occupying process, such as a tumor or metastasis, can trigger the same symptoms of course, and must be considered in the diagnosis. It is always difficult to solve the clinical problem of whether the symptoms are the result of a narrow spinal canal or a coincidental finding along with a systemic disease such as multiple sclerosis or amyotrophic lateral sclerosis.

Of pathophysiological significance is a complex, individual combination of damage from mechanical action and tissue ischemia. In the protective membranes such as the arachnoid mater, irritation leads to proliferation of connective tissue that forms a cuff around the spinal cord and constricts it. Constant microtraumas in the spinal cord also lead to damage to neurons and axonal transport, also causing demyelination and scarring. Circulation disorders in the spinal cord can be caused by compression. Bone ridges in the spinal canal can compress the anterior spinal artery or the pressure they exert on the spinal cord leads to circumscribed ischemia that plays a major role in the damage mechanism and can cause irreversible damage through microinsults.

The spinal cord that is fixated in the spinal canal by various ligaments has little room in the narrow canal for evasive actions (Fig. 32a,b). But they are constantly needed, as there is incessant movement at the spinal cord. This movement includes the transverse pulse or longitudinal respiration movements and the shear movement between bone and spinal cord that is triggered by every routine movement of the cervical spine. If hard structures such as bony spurs project into the spinal canal, the cord is particularly exposed to friction and pressure in that region.

Clinically, the signs of central spinal cord damage are in the foreground. For example, in the reflex test, enhanced proprioceptive reflexes with clearly widened reflexogenic zones are conspicuous, whereby the Babinski sign is positive. These indications of pyramidal tract damage explain the increased muscle tone up to spasticity, affecting primarily the legs in the form of paraspasticity. As the symptoms usually begin only discretely, attention must be directed to looking for fine motor disorders. For example, patients complain of problems unusual for them such as restraints when walking quickly and having muscles tire easily. Paraplegic sensory symptoms can ascend slowly from the feet to the neck, but pain is atypical. Occasionally there is dissociated dysesthesia. The irritation of central vegetative tracts can lead to bladder disorders; ultrasound examinations quickly detect subjectively unnoticed residual urine in the bladder. But bladder and rectal disorders are not particularly pronounced. In a classical case, signs of radicular lesion are found in the seg-

4

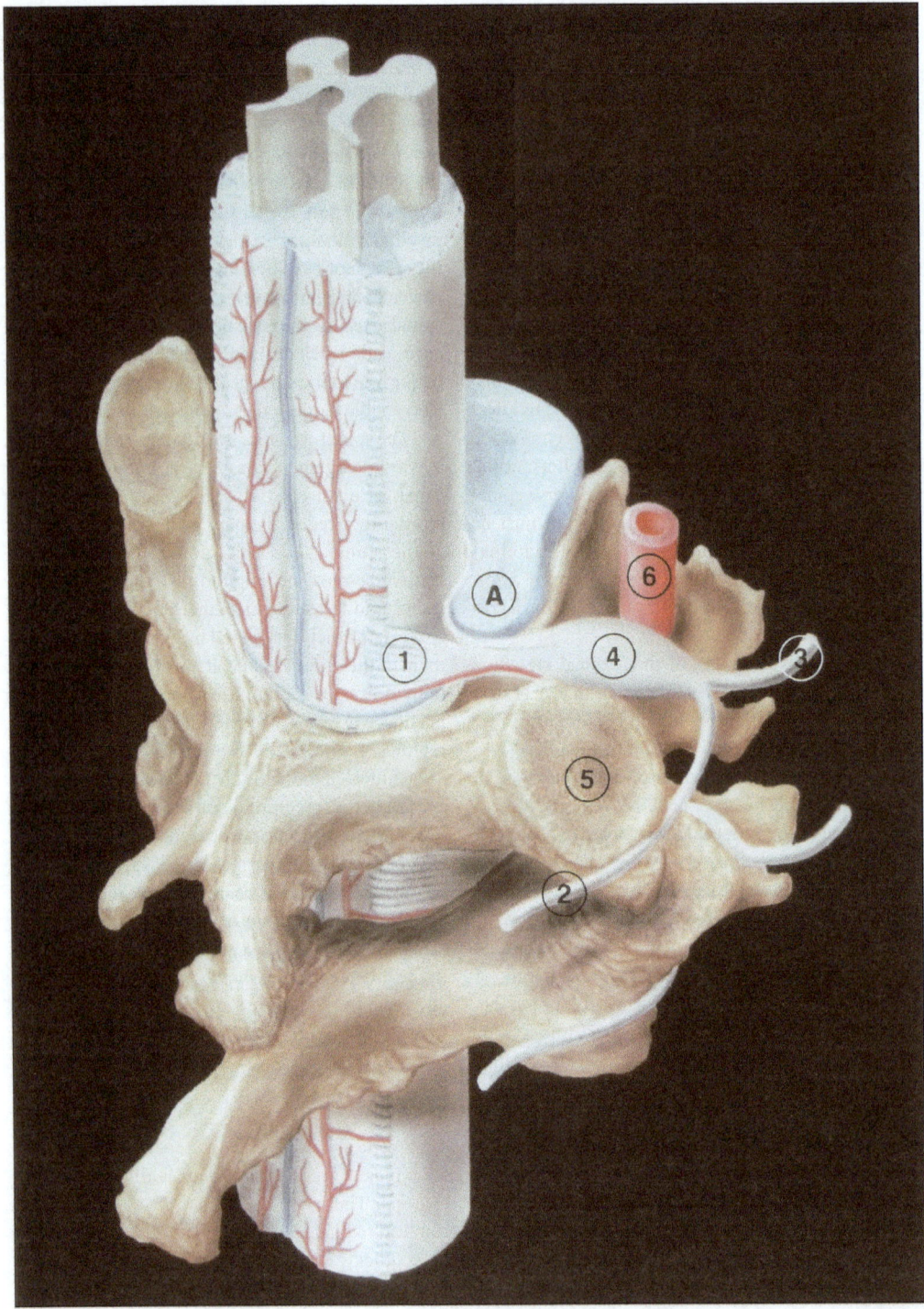

Figure 28 Effect of a laterally herniated disc on the root. 1 Dorsal root with the radicular artery, 2 Dorsal ramus of the spinal nerve, 3 Ventral ramus, 4 Spinal ganglion, 5 Spinal facet joint, 6 Vertebral artery, A Herniated disc compressing the ventral root

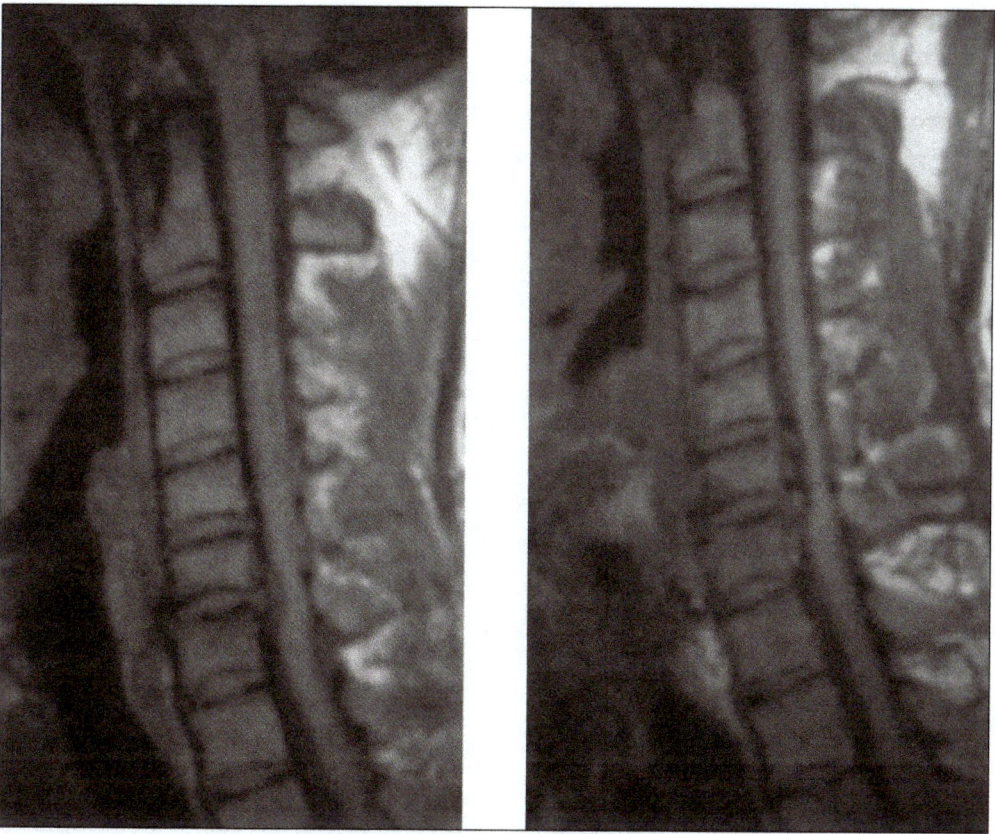

Figure 29 Magnetic resonance imaging. MRI of the cervical spine and the spinal canal. The two parallel sagittal planes show the spinal cord clearly compressed by prolapsed disc tissue at C6 and C7. Sequester from the prolapse is distributed over two segments. At the time of examination, the patient had already suffered spastic tetraparesis for several weeks with slight paresis and atrophy of the musculature innervated by C6/7

ment of the actual spinal cord lesion. They arise either from damage to the root or, if there are isolated motor deficits, from ischemic damage to the cells of the dorsal horn. About 30% of patients show pain syndromes with radicular character.

The clinical suspicion of a cervical myelopathy must always be clarified by radiological measures; today MRI is the standard procedure. Cervical myelo-CT has proved to be valuable for differentiated examinations, in particular if the CT or MRI have not answered all questions. In x-ray examinations of the cervical spine, cervical myelopathies are probable if the sagittal diameter of the spinal canal is less than 13 mm, and if it is under 10 mm they are extremely likely.

Stenoses of the spinal canal can also occur in the lumbar region. There, mainly circulation disorders of the cauda trigger pain and muscle weakness, occasionally also intermittent polyneuropathy-like sensory deficits with loss of reflexes.

For cervical myelopathy, the only useful therapy is releasing the spinal cord from its strictures. Surgery to remove circumscribed space-occupying masses such as herniated discs requires different access than releasing the spinal cord from connective tissue strictures. Long stretches of degenerative changes over several segments can often not be removed by surgery. To create space, the vertebral arches are removed from several segments and simultaneously, the denticulate ligaments and other

4

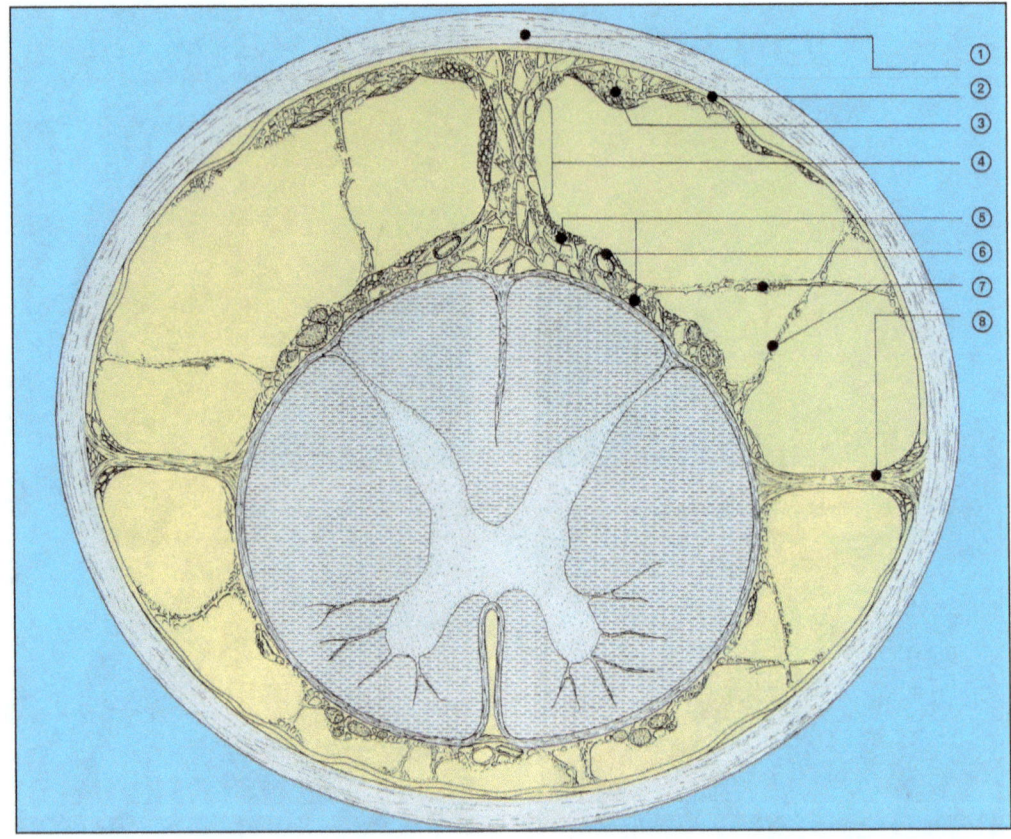

Figure 30 Spinal cord, dura, and connective tissue. The spinal cord is suspended in the spinal canal by ligaments. 1 Dura, 2 Arachnoid membranes, 3 Attachment of the septum and extent of the arachnoidea, 4 Connective tissue septum for fixating the spinal cord, 5 Epipial subarachnoidal tissue, 6 Vessels in the epipial network, 7 Trabecular membrane, 8 Denticulate ligament

strands of connective tissue are detached from the spinal cord (Figure 30).

4.7   Lumbar root syndromes

Herniated discs are more frequent in the lumbar spine than in the cervical spine; segments L4/L5 and L5/S1 are particularly affected. The acute event is often preceded by lumbago and sciatic pain, which already indicate a disorder in the functional segment. Most often affected are the middle aged (46–55 years), men more than women (Table 6).

The patient history is indicative for the diagnosis. If there is pre-existing disc degeneration, lifting a heavy object or an awkward rotation can trigger

the event. Radicular lesions usually occur with severe radicular pain radiating into the leg; numbness is also a frequent symptom. Patients note motor damage to the dorsal flexor of the foot as well as paresis of the quadriceps muscle when climbing stairs. Slight paresis of the calf muscles in S1 lesions cannot be assessed well in the examination, especially if testing is done against manual resistance. It is preferable to have the patient lift his own body weight by hopping on one foot. Similarly, the quadriceps muscle is tested by slowly climbing onto a chair. In the examination, a targeted search for symptoms of radicular irritation should be made. Symptoms of irritation can be triggered at the root mechanically by coughing or sneezing. Irritation can also arise during abdominal pressure due to a

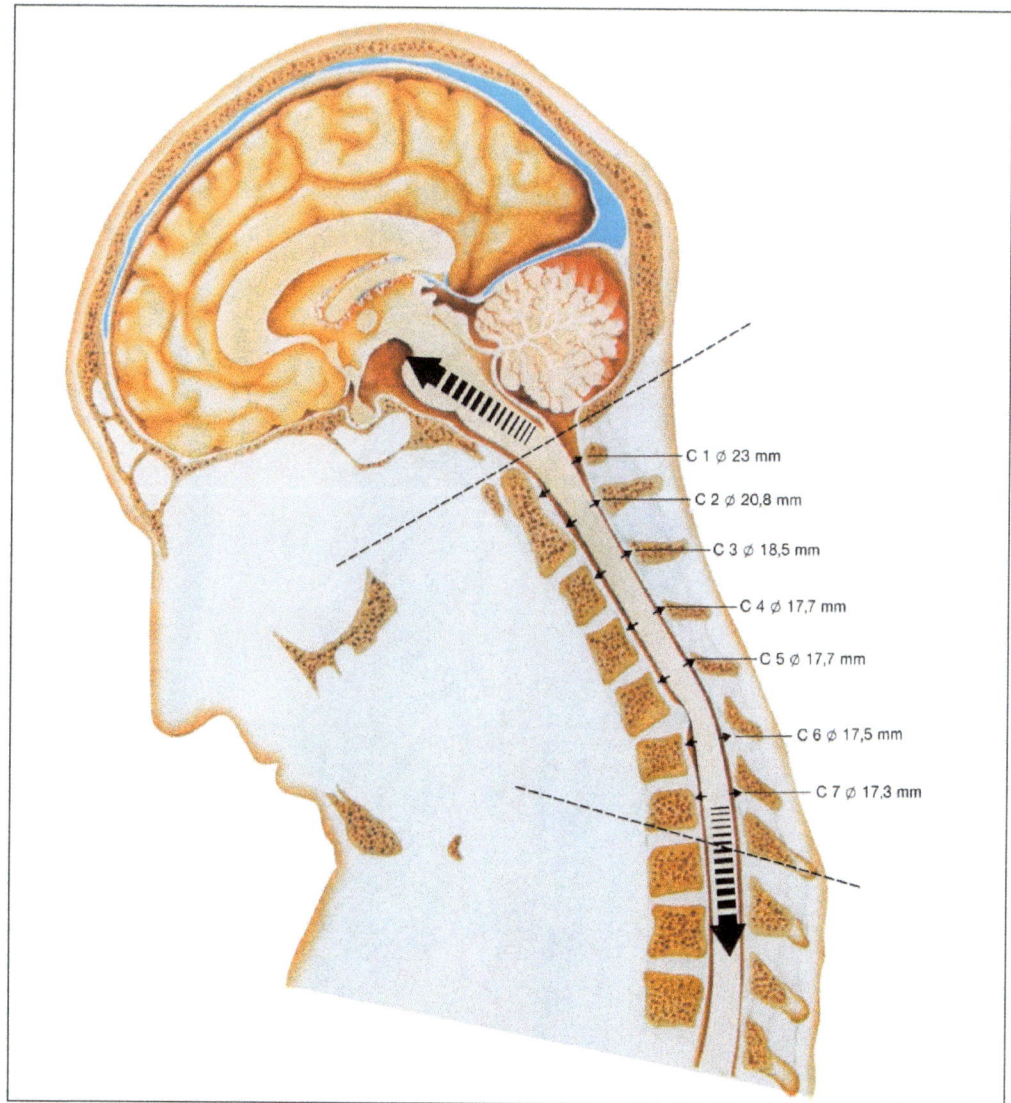

Figure 31 Functional anatomy of the cervical spine. Cervical spine with varying diameters of the spinal canal at the various segments. The diameter of the spinal canal is the shortest distance between the dorsal edge of the vertebra and the tip of the spinous process opposite it. Due to degenerative changes to the spine, pressure is put on the spinal cord or its physiological direction of movement (arrows) is obstructed

decrease in venous outflow which causes an increase in pressure at the root sheath and is immediately perceived as pain. Mechanical irritation of the sacral root can be triggered by the straight-leg raising test (Lasègue's sign). The outstretched leg of a patient lying on his back is lifted at the hip joint. The movement at the sciatic nerve is transferred as tension to the plexus and roots, triggering pain. If there is any uncertainty as to the veracity of the information, the patient should be later motivated to sit on the examination couch with outstretched legs. If he is able to do so without pain, his indication of pain is to be treated with skepticism. Testing the reverse Lasègue's sign is done by passive overextension of the hip joint

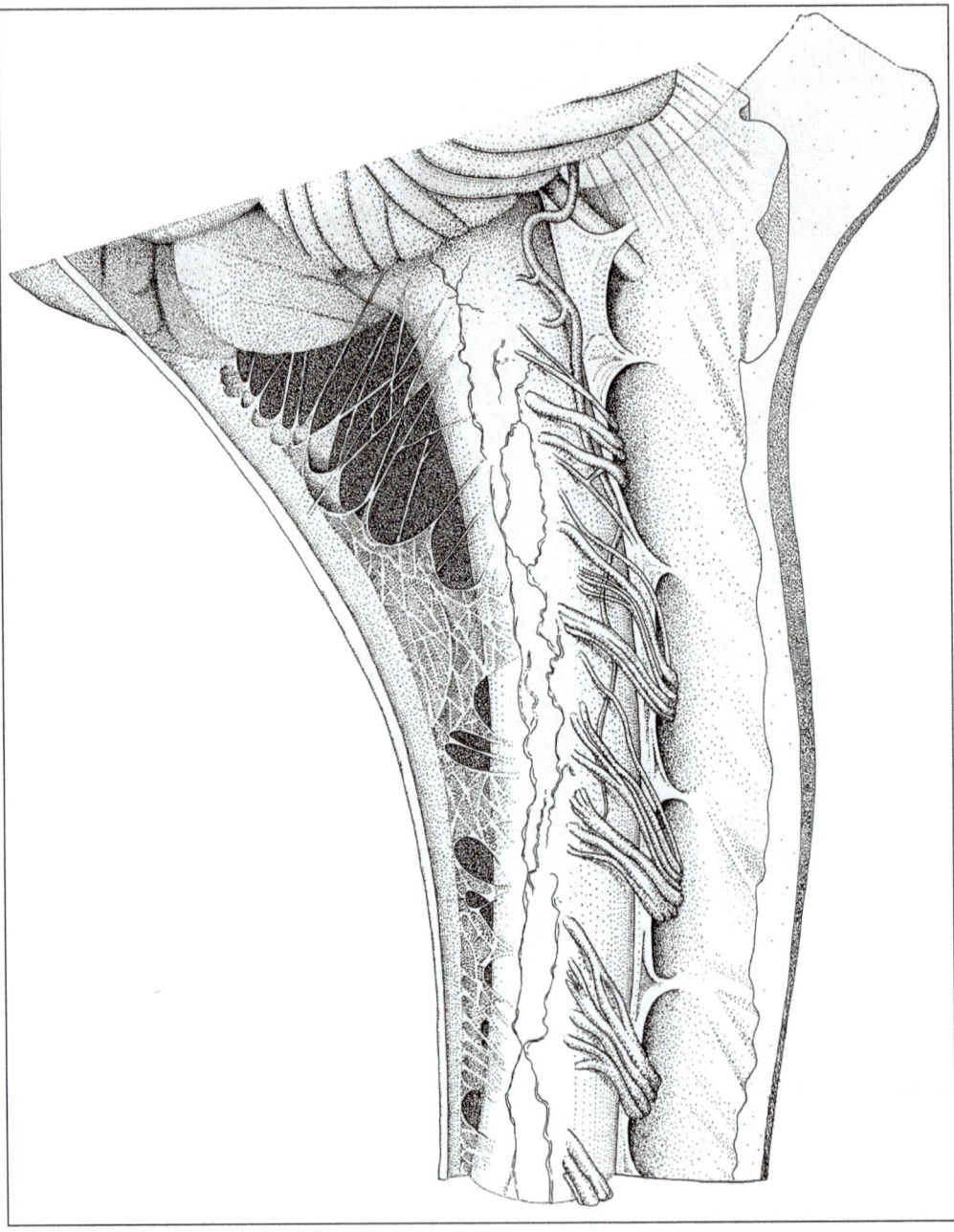

Figure 32 Graph of spinal cord and roots. The anatomy of the spinal canal is depicted in a transverse horizontal section in Figure 30. Image of spinal cord and root. The operation site shows the spinal cord and a spinal root with a dense vascular system overlay. Above and below the root, the sickle-shaped denticulate ligaments attach to the wall of the spinal cord. The structure of the denticulate ligaments as a perforated system of leaves enables the spine with the roots exiting from it to move unobstructed upward and downward over the necessary distance. The individual root fibers exiting from the spinal cord, the vessels, the denticulate ligaments, and the arachnoid membranes stretched between the spinal cord and dura in the operation site are visible in Figure 32a. The image shows the junction of the medulla oblongata to the cervical spinal cord with basal segments of the cerebellum

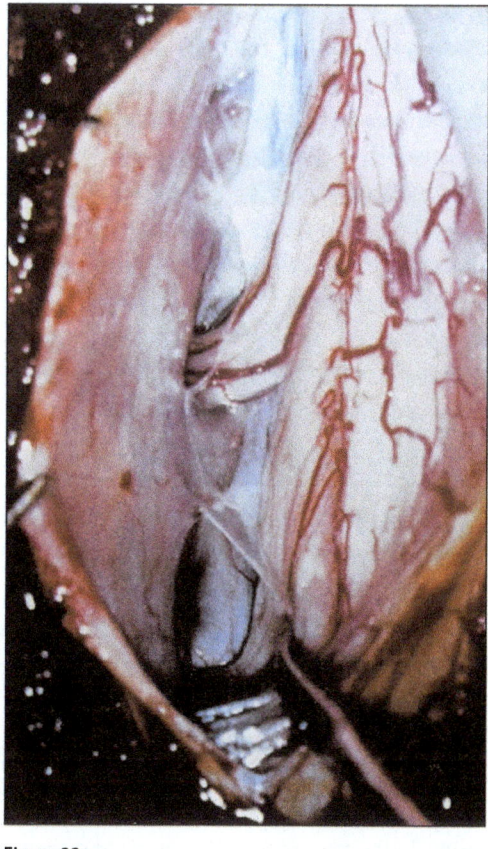

Figure 32a

and simultaneous flexing of the knee in a patient lying on his side (Figure 33). A positive deficit is found in lumbar root irritations at a higher level (L3/L4), but also in meralgia paresthetica and especially in arthrosis of the hip joint and lesions of the femoral nerve.

The physical examination reveals awkward posture with compensatory scoliosis of the spine; the paravertebral musculature is often very hard and tender to pressure, with extremely limited movement of the lumbar region. The examination shows the typical segmental motor or sensory deficits for a radicular lesion. However, in the lumbar region, mass prolapses frequently occur, which stem either from two segments or the sequester is so large that it damages two roots. An L4/L5 lesion can thus appear to be paresis of the peroneal nerve. In an L3 syndrome, the differential diagnosis must also distinguish paresis of the quadriceps muscle from damage to the femoral nerve.

4.7.1 Conus/cauda lesion

These symptoms can also be caused by a herniated disc that affects the end of the spinal cord, the conus medullaris (Figure 34).

Table 6 Summary of the lumbar root syndromes

Root	Sensitivity Pain	Indicator muscle Paresis Athrophies	Reflexes Disorders	Differential Diagnosis
L1	Pain in hip and groin			
L2	Front side of thigh	Psoas muscle		
L3	Dermatome from the head of the hip via the back of thigh to inside and knee	Quadriceps muscle	Patella reflex	Lesion of femoral nerve
L4	Dermatome on outside of thigh to patella and inside of lower leg to ankle	Quadriceps muscle Anterior tibial muscle	Patella reflex	Lesion of femoral nerve
L5	Dermatome above the knee, lateral on the lower leg moving downward to the top of the foot to the large toe	Paresis of the extensor hallux longus muscle, often also of the extensor digitorum brevis, gluteus medius, tibialis posterior muscles	Tibialis posterior reflex	Paresis of peroneal nerve
S1	Dermatome at the back of the thigh above the calf to the outer edge of the foot and small toe	Peroneal muscles, calf musculature, gluteus maximus	Achilles reflex	

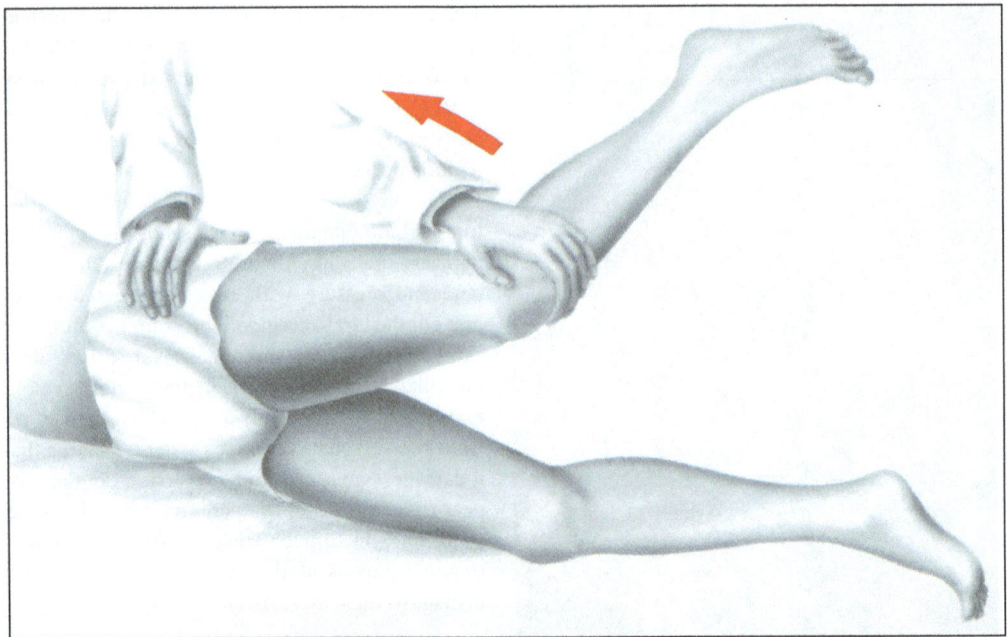

Figure 33 The inverse straight-leg raising test (reverse Lasègue)

Conus or cauda symptoms triggered by a prolapse are among the most urgent indications for neurosurgery. Every delay of the operation increases the probability that the neurological functions, in particular bladder, rectal, and sexual functions, will be irreversibly damaged. These functions are regulated by a complicated interaction of excitatory and inhibitory autonomous factors. It is important that the urogenital tract is initially controlled by autonomous parasympathetic innervation. For this, axons arise from preganglionic parasympathetic neurons of the lateral region of the sacral ventral horn S1–S4. They proceed outwards via the ventral roots and form the pelvic splanchnic nerves. Activation of these nerves mediates peristalsis as well as emptying of the urinary bladder and erection of the penis and clitoris. The sympathetic fibers proceed via the lumbar and sacral roots with somatic fibers from S2 to S4, which form the pudendal nerve. They regulate the external sphincter of the bladder, the lower rectum and anus, and control sexual functions.

Complete symptomatic cauda causes an autonomous bladder. Due to a lack of bladder tone, the organ is limp and distended, and the feeling of a full bladder is lacking. This leads to maximum filling of the bladder and corresponding amounts of residual urine. In the stage of the overflow bladder, there is incontinence with a constant dribbling of urine and danger of ascending urinary tract infections.

The only correct medical decision for acute conus-cauda lesions is immediate referral to a hospital that is diagnostically and surgically capable of treating such emergencies.

4.7.2   Spondylolisthesis

Spondylolisthesis is an anteroposterior translatory movement of two spinal vertebrae in relationship to each other caused by instability between the two involved vertebrae (Figure 35). When the upper vertebra glides over the lower one, an x-ray shows a clear step in the region of the posterior vertebral edges. This congenital anomaly occurs most frequently in segment L5/S1. When the vertebral joints of the two vertebrae show interrupted continuity, spondylolysis exists. The instability can be caused

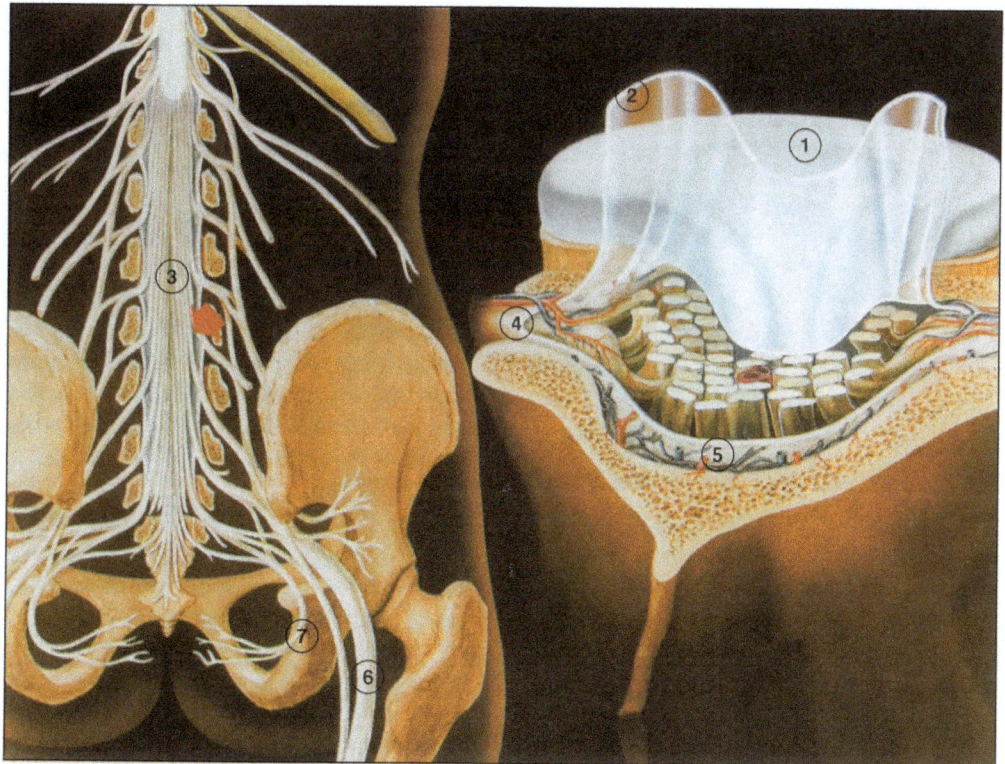

Figure 34 Conus/cauda lesion. (1) Medial prolapse of the nucleus pulposus, which presses against the dura (2) and compresses the fibers of the cauda equina (3). The spinal root [(4), red arrow], which is usually damaged in a lateral herniated disc, is free in the foramen. The vascular supply of the roots, which is usually damaged in root compression, can be easily seen; (5) epidural space with venous plexus; (6) sciatic nerve; (7) pudendal nerve. The massive lumbar medial prolapse can constrict the spinal canal so much at any level that several roots, right and left, are displaced and squeezed against the wall of the spinal canal. This causes transverse damage to the cauda equina. Since the sacral roots are closest to the midline, they are usually the first and most affected. Clinically, there is often paralysis of the calf musculature on both sides, reflex deficits on both sides, and a sensory disorder known as saddle block hypoesthesia to anesthesia, referring to its location on the parts of the body that would be in contact with a saddle. There is always a disorder of bladder, rectal, and sexual functions

by degenerative changes of the facet joints, or by traumatic or congenital lesions of the pars interarticularis of the upper of the two vertebrae.

In this genetically caused disorder, the displacement of the vertebra occurs during the growth phase and then stops. Disc damage often occurs in this disease from mechanical degeneration. However, the pain is less often caused by disc herniation than from static and mechanical degeneration. Spondylolisthesis can be a coincidental finding in an x-ray; it need not necessarily cause clinical deficits. In many cases, the distortion of the anatomy induces mechanical compression of one or more lumbar nerve roots. This can cause sciatica or disturb bladder or bowel function. If the diameter of the spinal canal is greatly decreased in severe cases, a cauda equina syndrome can be induced. The efficacy of fusion surgery in addition to decompression surgery in patients who have lumbar spinal stenosis with or without degenerative spondylolisthesis has recently been studies. The decompression surgery plus fusion surgery did not result in better clinical outcomes at 2 years and 5 years than did decompression surgery alone. As decompression is the easier treatment, this surgery should be performed (Forsth P, Olafsson G, Carlsson T, et al. N Engl J Med 2016;374:1413-23).

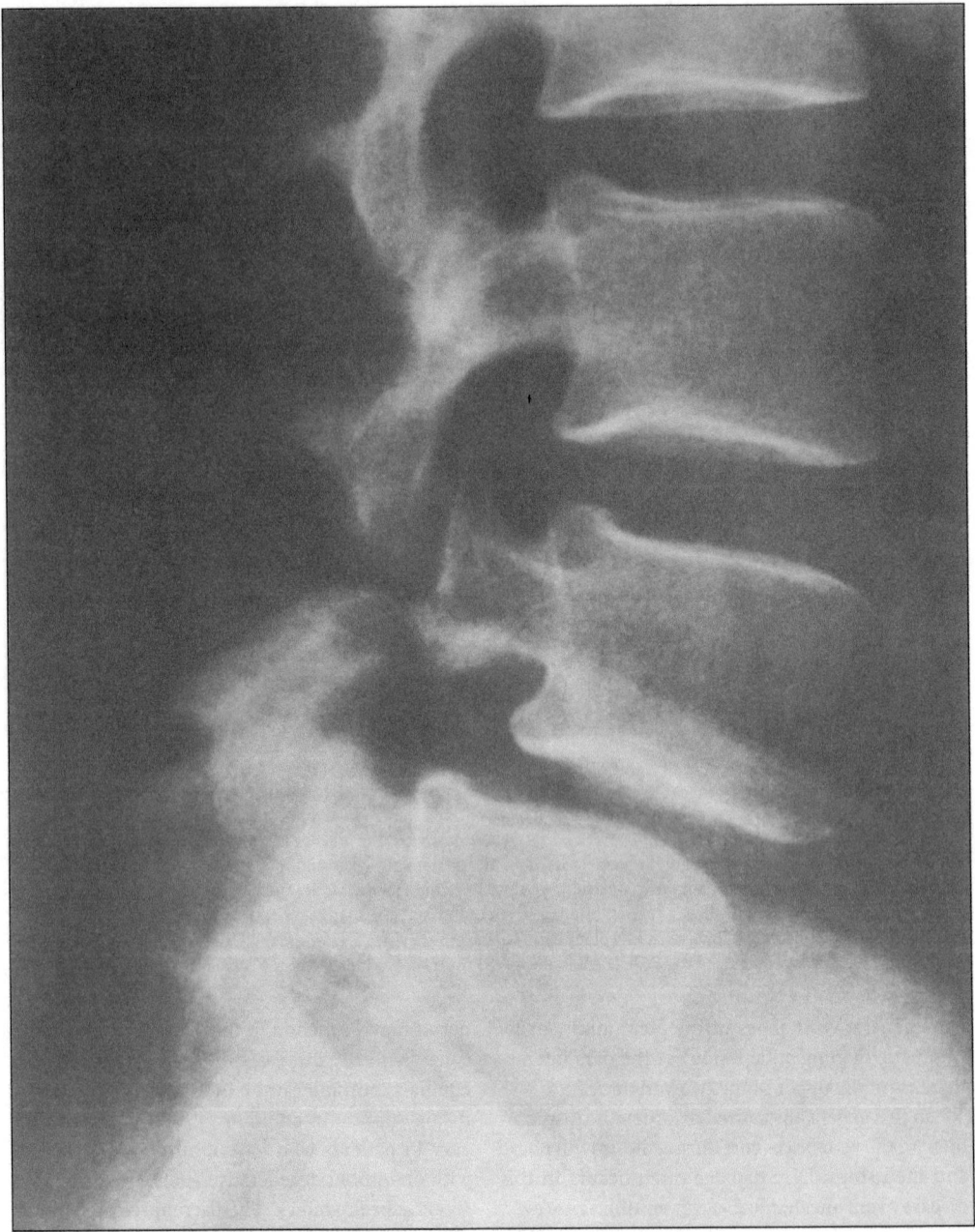

Figure 35 X-ray finding of spondylolysis. The x-ray image shows spondylolysis. It is characterized by a vertebra slipping forward. The 4th lumbar vertebra is displaced ventraly to the 5th. Due to the formation of a gap, the vertebra with its vertebral arch and upper facet joints is separated from the posterior part of the vertebral arch with the lower joint processes

Lesions of single nerves

5.1 Trigeminal nerve

The trigeminal nerve exits the brain in its anatomical course in the area of the pons and proceeds to the anterior edge of the petrous bone. There, the nerve forms the trigeminal ganglion – also known as the Gasserian ganglion – which, sheathed by a dural sleeve, gives off three branches (V1–3). These branches are responsible for sensory supply in the face and form three innervation zones corresponding to the supply territories of the ophthalmic nerve (V1), the maxillary nerve (V2), and the mandibular nerve (V3). Damage to the nerve in its peripheral course results in deficits in the bands of innervation zones.

Central damage leads to a different picture. Damage to the large stretch of the sensory trigeminal nuclei in the brain stem leads to circular, onionskin-like layers of sensory deficits in the face. Lesions in the upper segment of the nucleus affect the nose and mouth region, damage located further caudal or deeper affects the facial areas extending back to the ear corresponding to the Laehr-Soelder lines.

5.1.1 Corneal reflex

The ophthalmic nerve innervates the cornea, where the corneal reflex can be triggered by stimulation with a cotton tip. This elicits involuntary blinking in both eyes. To do so, the efferent reflex arc in the facial nerve must naturally be intact. The afferent reflex arc (V1) can be damaged in the peripheral segment by a zoster infection. In trigeminal neuralgia, however, the reflex is normal. If there is a deficit, it requires urgent further investigation. The greatest significance of the reflex is to determine damage to the brain stem, in particular in comatose patients. An absent corneal reflex indicates severe damage to the brain stem.

5.1.2 Sense of smell

The nasal mucosa are innervated by branches of the trigeminal nerve, the nasociliary nerve (V1), and the pterygopalatine nerve (V2). They react to ammonium chloride, which is a trigeminal irritant and

in the past was filled into vials and used as smelling salts to revive people who had fainted. The reaction to ammonium chloride should always be tested if, in a test of the sense of smell, aromatic substances cannot be perceived via the 1st cranial nerve, the olfactory nerve. Lack of a reaction to ammonium chloride indicates either severe damage to the nasal mucosa or a psychogenic disorder of the sense of smell.

5.1.3 Bulbar and pseudobulbar paralysis

Paralysis of the voluntary motor functions in the facial area requires a very precise analysis as to whether individual cranial nerves, certain groups of cranial nerves, or in systemic diseases (amyotrophic lateral sclerosis, polyradiculitis, or polyneuropathy) all motor nuclei or their axons are affected. Anatomically, the most important motor fibers of the trigeminal nerve proceed via the masticator nerve, which branches off from the V3 and supplies important chewing muscles such as the masseter muscle, temporalis muscle, and pterygoid muscles. Damage to the nerve is rare and is made apparent by a displacement of the jaw to the paralyzed side when the mouth is opened. The pronounced asymmetry of the corners of the mouth occasionally found with paresis of the facial nerve can appear to be displacement of the jaw and lead to a misdiagnosis. Bilateral weakness of the chewing musculature occurs either in bulbar or pseudobulbar paralysis. In bulbar paralysis, the motor nuclei are damaged, the cranial nerve is degenerated, and the musculature shows spontaneous EMG activity. In pseudobulbar paralysis, however, the pyramidal tract is damaged and the intact nuclei of the cranial nerve cannot be voluntarily activated from the central system. The musculature develops inactivity atrophy; there is no spontaneous EMG activity.

5.1.4 Masseter reflex

The masseter reflex, which is elicited by tapping the chin while the mouth is loosely and slightly opened, is very significant. The reflex tract of this proprio-

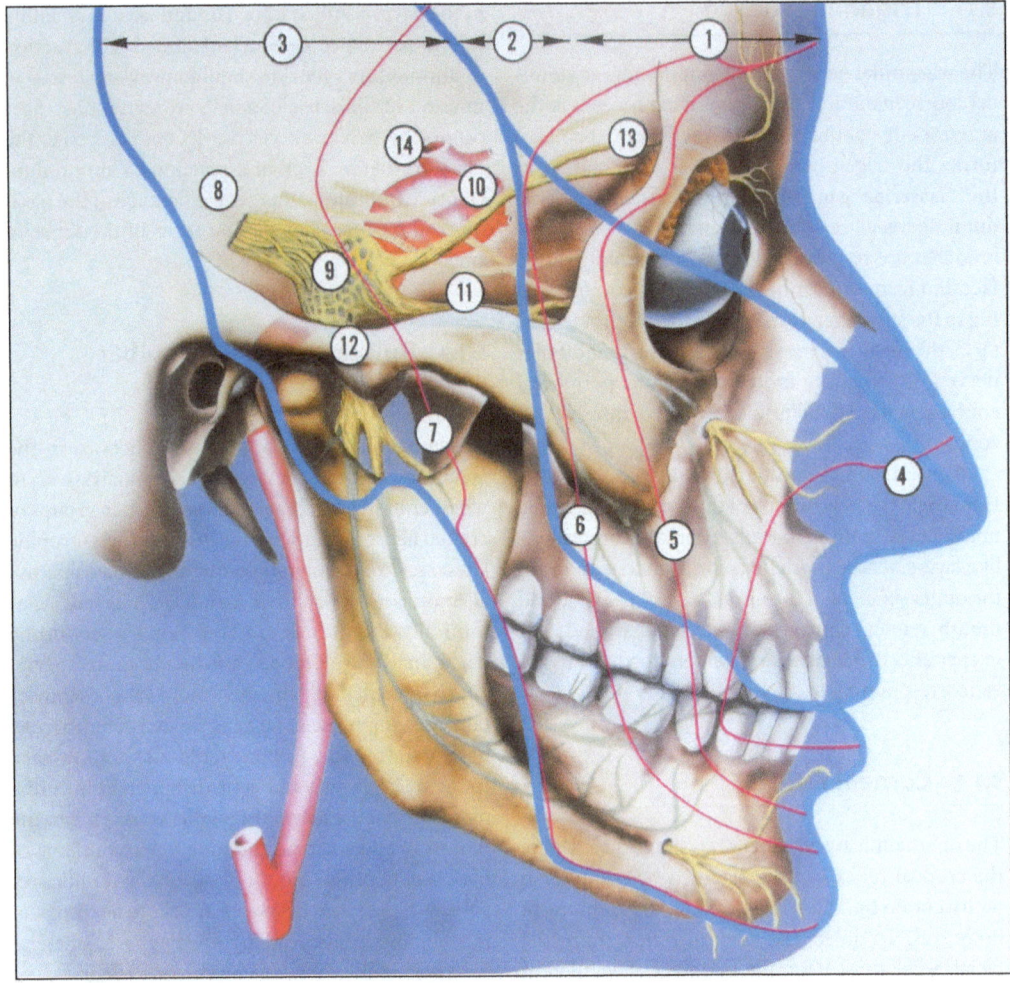

Figure 36 Diagram illustrating trigeminal nerve anatomy. 1 Area supplied by the ophthalmic nerve, 2 Area supplied by the maxillary nerve, 3 Area supplied by the mandibular nerve, 4 The Laehr-Soelder lines show the "onion skin-like" construction of the trigeminal spinal nucleus in its representation of the face's sensory skin innervations. 4–6 represent rostral areas, 7 Dorsal area, 8 Trigeminal nerve, 9 Semilunar ganglion of the trigeminal nerve, 10 Ophthalmic nerve, 11 Maxillary nerve, 12 Mandibular nerve, 13 Lacrimal nerve, 14 Internal carotid artery with aneurysmatic malformation

ceptive reflex proceeds via sensory trigeminal fibers to the brain stem and from there via the motor trigeminal nucleus and the masticator nerve to the masseter and temporal muscles. An enhanced masseter reflex indicates a lesion of the pyramidal tract located in the brain stem or higher. However, if the masseter reflex is normal, but the other proprioceptive reflexes on the extremities are enhanced, it can be concluded that there is a lesion in the lower brain stem region down to the upper cervical spinal cord.

The reflex is thus a valuable diagnostic aid for determining the level of damage in this region.

5.1.5 Trigeminal neuralgia

Clinical symptoms

Trigeminal neuralgia is the most common problem with this nerve and the most frequent cause of facial neuralgia (Figure 36). It causes intense, fulgurant, one-sided pain with a tearing, burning sensation, lasting seconds at the most, rarely longer, and less than two minutes. The face is distorted in pain when an attack occurs. This gave the disease its name tic douloureux. Often, the pain attacks can be repeated in intervals of only minutes. There is no pain in the intervals. Women suffer more frequently from this neuralgia than men; the right side of the face is affected twice as often as the left side. In almost 40% of cases, the pain is located in the area supplied by the 2nd and 3rd branch. Because the pain occurs in the area of the upper or lower jaw or teeth, a dentist is frequently consulted. This leads to many unnecessary tooth extractions. Short-term pain, initially only twinges with an increasingly paroxysmal character, can precede the classic symptoms for a few months and should be treated as the full-blown illness. Neuralgia pain usually occurs spontaneously with no warning. However, in the examination a trigger zone is always found. It consists of a small area of skin, which when touched, triggers the attack immediately. It is often a small area around the nasolabial fold that can be irritated by the slightest facial movements, causing an attack. Eliciting factors can be laughing, tooth brushing, swallowing, and sometimes even shaking from walking or even a breeze. As eating in particular is a trigger, pronounced weight loss is an indication of the severity of the illness.

Diagnostics

Space-occupying processes of the posterior cranial fossa or at the sphenoid wing can manifest themselves as symptoms of trigeminal neuralgia. Therefore neurinomas, meningeomas, aneurysms, and metastases in the visceral cranium must be ruled out by CT or MRI of the base of the skull and petrous bone. A precise patient history and eliciting a concrete description of the pain from the patient as well as the neurological examination provide indications as to the existence of intracerebral diseases such as multiple sclerosis or a brain stem process which need to be further investigated.

Surgical reports indicate that intraoperatively, a pathological vessel-nerve contact was found in 70–100% of patients. The attempt to reveal vascular compression preoperatively through MRI depends on the MRI technique used. Sensitivity of up to 88%

can be achieved, but the specificity is only about 50%, as pathological vessel-nerve contacts can also be found in about a quarter of control subjects.

Therapy

In Chapter 7 "Pain mechanics," various pathomechanisms for the development of pain are named that apply to facial and trigeminal neuralgia. This pathophysiology determines the treatment approach. In the discussion of the etiology of neuralgia, it is interesting to note that the age group of 50–80 year olds is particularly affected. It is unclear why younger persons are not affected more often. Possibly there is a time factor that allows the triggering causes to become effective only at an advanced age. Degenerative processes at the vessels lead to elongation and coiling, as is often observed in the temporal artery in older persons. This concept assumes that vascular coils throb due to the pulse against the central nerve stem, causing microtrauma damage to the myelin. Nerve compression is most frequently caused by contact with the superior cerebellar artery (about 80%), less frequently and in decreasing order with pontine veins, the inferior anterior cerebellar artery, or other smaller vessels.

An end of the pain caused by this can often be achieved by surgical padding of the nerve (vascular decompression). In addition, demyelinating diseases such as multiple sclerosis – in the nerve or central – can interrupt afferent tracts or stimulate central neurons to increase activity. Since these persons are usually younger than 50, demyelinating disease must be ruled out in these cases. To avoid unnecessary confusion it is useful to know that the first manifestation of multiple sclerosis is extremely unlikely in persons over 50.

Initially, conservative therapy should be adequate and specific. Most important is therapy with antiepileptics (Table 7). Only therapy-resistant neuralgias should be operated.

Carbamazepine is the treatment drug of choice. Among long-term medication, it has the fewest side effects, but also the disadvantage that is can be administered only orally. If a therapeutic level is to be reached quickly with oral medication, it can usually only be done with side effects that can be very unpleasant. The patient must be thoroughly informed of these. Depending on the acuteness of the symp-

Table 7 Drugs that have proven to be effective in treating trigeminal neuralgia

Drug	Starting dose	Therapeutic level (max. dose)
Carba-mazepine	100–200 mg	600–1200 mg (1400 mg)
Oxcarba-mazepine	300 mg	600–1200 mg (2400 mg)
Lamotrigine	25 mg	100–200 mg (400 mg)
Baclofen	5 mg	20–80 mg (80 mg)
Pregabalin	75 mg	150 mg (600 mg)

toms, at the beginning, 100–400 mg twice daily should be administered, according to the level of individual tolerance. The dosage can then be increased by 50 mg daily until the patient is pain free or intolerable side effects occur. Due to the great differences in tolerability, this procedure is better than therapy control by checking plasma levels. When a satisfactory therapeutic dosage is reached, a maintenance dosage for several weeks with 300 mg 2 to 3 times a day with a depot compound is recommended. Known side effects at the beginning of treatment are vertigo, ataxia, and double vision. For immediate therapy, if urgently needed, 250 mg parenteral phenytoin can be given cautiously by intravenous or infusion concentrate (750 mg diluted in 50 ml NaCl) over 30 min. Further saturation of phenytoin can follow as needed intravenously or orally (3 mg/kg body weight in 3 doses).

Another medication, the antispastic drug baclofen, has also proven effective. This substance can be given as monotherapy or in addition as co-medication with carbamazepine, in particular if the latter was not sufficiently effective despite a high dosage. Initially, 10 mg of baclofen are given, which can be increased up to 3 × 10 mg and occasionally more.

The administration of analgesics is not effective for treating neuralgia. Neuroleptics are effective only as adjuvant medication due to their sedating component. A positive treatment effect is also reported from gabapentin, pregabalin and lamotrigine.

Surgical measures extend from vascular decompression or thermocoagulation up to gangliolysis through alcohol or glycerin injection. Studies on the immediate effect of decompression surgery report a pain-free state in 82% and reduced pain in 16% of patients. But this effect is not long lasting; after 10 years, only 53% of the patients still report being pain free. If a second decompression surgery became necessary, positive results were achieved in only 51% of the patients. All measures can lead to freedom from pain. In addition to the complications of surgery itself, it must be considered that despite the operation, the neuralgia frequently reoccurs a few years later and must then be treated again. The rate of surgical complications increases and the success rate regarding freedom from pain is reduced. All of this emphasizes the necessity of first exploring the limits of conservative therapy professionally and systematically.

Differential diagnosis of trigeminal neuralgia
Costen's syndrome

Pain that radiates into the head, is localized in front of the ear, and extends to the jaw can be caused by poor occlusion and functional disorders of the jaw. In conjunction with habitual teeth grinding, it causes irritation in the mandibular joint that becomes sensitive to pressure. Costen's syndrome does not manifest itself as attacks, but more as continuous pain and requires dental treatment. Orthodontic measures relieve muscle strain and the projected pain disappears.

Glossopharyngeal neuralgia

This rare neuralgia is marked by one-sided, fulgurant pain into the tonsil region or base of the tongue. Swallowing and drinking cold beverages irritates triggering points in the throat and causes the pain attacks.

5.2   Facial nerve

5.2.1   Anatomy

From the nucleus of the facial nerve at the floor of the 4th ventricle, the axons proceed in an arc around the nucleus of the abducens nerve and exit the brain in the cerebellopontine angle behind the trigeminal

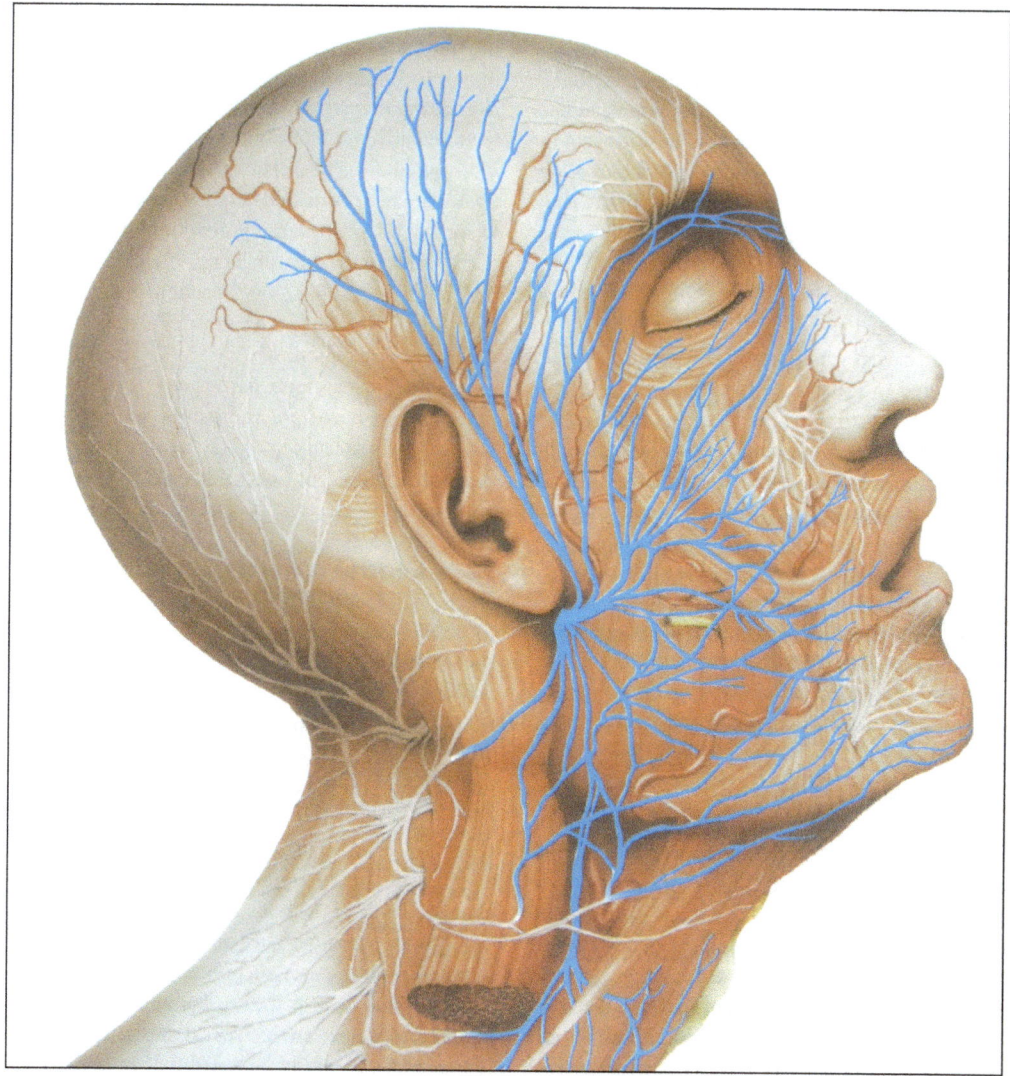

Figure 37 Schematic diagram of the facial nerve. The motor facial nerve exits it bony canal under the ear and innervates the muscles for facial expression. In front of the ear, it crosses the salivary gland, of which only the canal (yellow) is visible here. In addition to the temporal artery, branches of the cervical plexus nerve and the sensory nerve exits of the trigeminal nerve above and below the eye and at the tip of the chin are shown

fibers. Both proceed into the internal ear canal close to the acoustic nerve and can be accompanied here by the anterior inferior cerebellar artery. In the internal acoustic pore, the facial nerve exits to its own 25- to 30-mm-long bony canal, the facial nerve canal. The nerve exits the canal in the stylomastoid foramen of the petrous bone and divides into several motor branches that supply the muscles for facial expression at the forehead and eyes, in the cheek and mouth area, and the platysma at the neck (Figure 37).

A few fibers also anastomize with motor segments of cervical nerves. The branching of the nerve at the parotid can easily lead to its being damaged there if inflammatory or tumorous changes affect the gland.

The geniculate ganglion, from which parasympathetic fibers branch out to supply the lacrimal gland and the mucous glands in the nose and throat, is located in the facial nerve canal. Somewhat later the stapedius nerve exits the canal to innervate the stapedius muscle, which controls tension of the tympanic membrane. If there is paresis of the facial nerve, paralysis of the stapedius muscle can cause hearing impairment (dysacusis, hyperacusis). In the lower part of the canal, the tympanic chord exits as a purely sensory nerve and proceeds through the tympanic cavity, where it and the motor nerve are endangered in the event of otitis or zoster oticus. Shortly after the lingual nerve, the tympanic chord attaches to the 3rd trigeminal branch to register the tastes sweet, sour, and salty in the front two thirds of the tongue and to innervate the sublingual and submandibular glands.

5.2.2 Peripheral facial nerve paresis

The peripheral facial nerve lesion is a disorder with many causes (Figure 38a). If no specific cause can be determined after a careful examination, it is an idiopathic paresis, which occurs most frequently. It is assumed that an inflammation causes edematous swelling of the contents of the facial canal. The increase in pressure associated with this leads to demyelination and, in severe cases, to axonal nerve damage. The severity of the damage pattern cannot be accurately assessed in the acute stage. However, this would be preferable, as the severity of the lesion determines the prognosis of the illness. Lesions also occur as concomitant symptoms of other illnesses. For example, tumors, inflammatory or infiltrative nerve involvement, or vascular loops in the cerebellopontine angle can cause lesions, and inflammatory (herpes oticus, borreliosis), tumorous (cholesteatoma) or mechanical damage (petrous bone fractures) can damage the nerve in the narrow bony canal, as can cuts or blunt trauma to the face.

The level of the lesion on the nerve determines the clinical pattern of deficits. The nerve is most frequently damaged in the canal, which causes a frozen facial expression on one side. The forehead cannot be wrinkled, the eyelid remains open, the paralysis causes the mouth to remain unmovable when speaking and laughing. The paralyzed side of the mouth "hangs," the philtrum at the upper lip is drawn to the healthy side, cheek tone and mouth closure are defective, when the teeth are bared, the teeth remain covered by the paretic lips. If the geniculate ganglion is also damaged, no tears flow, the open eye can dry out, and keratitis can destroy sight. Disorders of taste or saliva production – in one third of those affected – are usually not discovered except by direct questioning or thorough examination. Many patients also report sensory deficits in the cheek and oral region.

Bilateral facial nerve paresis can occur in trauma, Guillain-Barré polyradiculitis, or tick-borne radiculitis, as well as with basal meningitis, meningeosis neoplastica, or Boeck's disease.

5.2.3 Central facial nerve paresis

In incomplete facial nerve paresis, the deficit pattern is more discrete and requires a thorough examination. In particular, central facial nerve paresis should be distinguished (Figure 38b).

Central facial nerve lesions can also lead to a one-sided paralysis of the face, but wrinkling the forehead is always possible and EMG finds no spontaneous activity in the paretic muscles. In addition, an increase in the proprioceptive reflexes on the ipsilateral side of the body indicates a central nervous system event with central paresis of the facial nerve. A central lesion, such as an insult in the brain stem, can also destroy the nucleus and intracerebral axons and cause an ipsilateral peripheral facial nerve lesion. It should be noted that in such brain stem insults, the proprioceptive reflexes on the contralateral side are enhanced, as the pyramidal tract does not cross over to the contralateral side of the body until later and deeper in the brain stem. Pathways leading to the nucleus of the facial nerve can be damaged in multiple sclerosis, tumors, or polyradiculitis and trigger facial nerve myokymia, continuous or intermittent, asynchronous muscle contractions or tics, usually periocular or perioral.

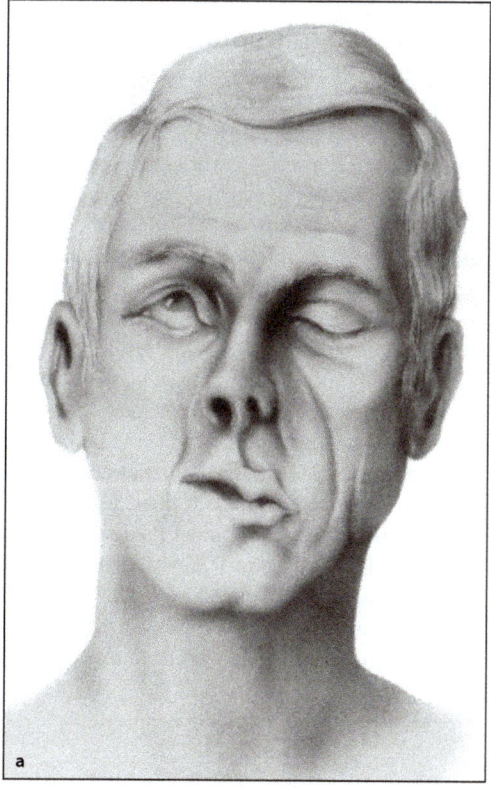
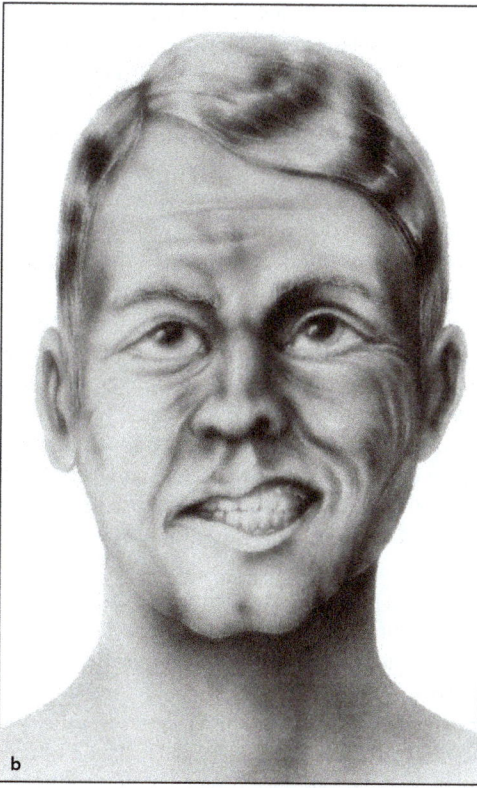

Figure 38 Facial nerve palsy. **a**: Right peripheral facial nerve palsy. The attempt to close the eyes is unsuccessful on the right side. The upward movement of the eye is known as Bell's phenomenon. When trying to whistle, the mouth is drawn to the left toward the healthy side. It is not possible to wrinkle the right forehead. **b**: Right central facial nerve palsy. When baring the teeth, the right corner of the mouth does not move. The forehead can be wrinkled on both sides

5.2.4 Facial nerve spasm

Here, pathomechanical events occur in motor nerves just as has been already described for sensory nerves in trigeminal neuralgia, in particular irritation from vascular loops. Nerve irritation leads to spontaneous ectopic and ephaptic neural activity that leads to severe synchronous contractions in the entire facial musculature. Emotional excitement with movement in the facial area can elicit a spasm attack. In addition to psychogenic disorders, a differential diagnosis of blepharospasm must be considered, which as an extrapyramidal dyskinetic syndrome can lead to spasmodic closure, usually of both eyes, and can also affect other muscles not innervated by the facial nerve in the sense of torticollis. Recently, therapeutic local in-jections of the affected muscles with botulinum toxin have proved to be useful for these organic diseases.

5.2.5 Therapy

Idiopathic facial nerve paresis develops progressively within hours up to 2 days to its fullblown symptoms and usually recedes in 4–6 weeks. There are thus far no convincing studies on idiopathic paresis that prove the superiority of medication or surgical therapy over just waiting. Without any therapy, healing with no or hardly any residual symptoms can be expected in 4 out of 5 patients. Many reports recommend giving cortisone (prednisone; 1st week 60 mg daily, 2nd week, gradually tapering off). This should cause the initial inflammatory pain to recede more quickly, especially if there is zoster oticus. In addition, the cortisone is said to have a positive effect on the prognosis regarding good functional reinnervation, with development of less synkinesis. Prednisone combined with B vitamins is reported to be effective, especially in combination with methylcobalamine. If there is a suspicion of a virus infection as cause of the inflammation, 2,000–2,400 mg acyclovir daily can be given for 10 days. But a double-blind, placebo-controlled study recently showed that there is no evidence for any benefit if acyclovir is given alone or any additional benefit of acyclovir in combination with prednisolone. A treatment of 25 mg prednisolone twice daily proved to be effective and superior to 5×400 mg acyclovir per day. The patient's expectations and the physician's varying motives for providing treatment often result in various polypragmatic treatment schemes, up to definitely unnecessary inpatient infusion therapy. The more severe the clinical deficits are, the greater is the probability that permanent damage in the form of paresis will remain. Here it is very important to protect the eyes; salves, eye dressing, or an eye patch should be used to prevent damage. Electrotherapy is not very helpful, as no isometric muscle contractions can be done. Instead, the patient should be instructed to carry out exercises on his own, such as practicing facial expressions in front of the mirror.

The pathological motor co-innervation or synkinesis that can be frequently observed in a thorough examination of facial nerve paralysis is treated in the Sect. 1.7, "Denervation and reinnervation." Faulty vegetative regeneration is the cause for gustatory sweating or crocodile tears. At the sight of food, an innervation disorder causes the lacrimal gland instead of the sublingual gland to be stimulated.

5.3   Brachial plexus

The brachial plexus (Figure 39) is often exposed to mechanical damage, in particular traumatic damage, due to its particular topographic location near the highly movable structures of the shoulder girdle. The exact determination of the location of the lesion is not always easy due to the complicated structure of the brachial plexus. The plexus gets inflows from the spinal roots, which contact each other at different levels of the plexus and then disentangle again as they approach the peripheral nerve. Depending on the location of the damage, shoulder girdle and arm muscles are affected in different combinations due to their plurisegmental innervation and varying supply from the plexus and peripheral nerve segments.

The brachial plexus receives its inflow from segments C5–T1. Upward or downward displacement by one segment (C4 and T2) is still within the norm. The ventral branches of the spinal nerves combine to three primary tracts, or trunks; the superior trunk arises from the union of C5 and C6, possibly with involvement of C4. The medial trunk arises from C7, and the inferior trunk is formed from C8 and T1. Each trunk has a ventral and a dorsal branch. These trunk branches form three fascicles, which typically group around the axillary artery. The arm nerves then arise from the fascicles (Figures 39, 40).

The arm nerves are formed from the following combinations:

Lateral fascicle	Musculocutaneous nerve (C5–C7)
	Median nerve (C5–C7)
Medial fascicle	Median nerve (C8–T1)
	Medial brachial cutaneous nerve (T1)
	Medial antebrachial cuta neous nerve (C8–T1)
Posterior fascicle	Axillary nerve (C5–C7)
	Radial nerve (C5–C8)

5.3.1   Traumatic damage

Upper plexus

In traumas, the uppermost segment of the plexus that is formed from roots C5 to occasionally C7 is most frequently affected. This results in the Erb-Duchenne type of paralysis. Clinically, sensory deficits are less apparent than motor deficits. Usually, shoulder joint movement is limited due to difficulties in abduction and outward rotation and paresis of the brachial biceps and supinator muscles. This paresis results in a characteristic arm position and is thus a useful diagnostic symptom. The hang-

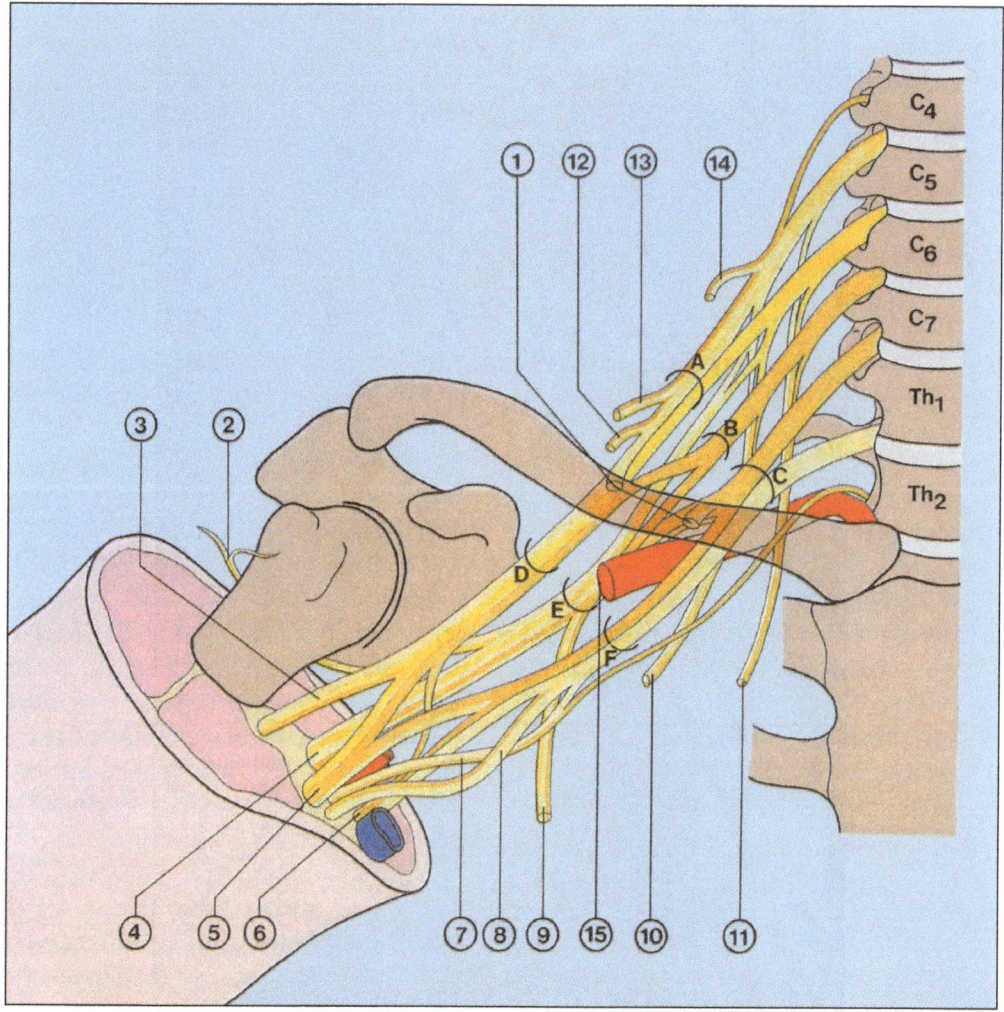

Figure 39 Scheme of the anatomy of the brachial plexus. 1 Nn. pectorales, 2 N. axillaris, 3 N. musculocutaneus, 4 N. radialis, 5 N. medianus, 6 N. ulnaris, 7 N. cutaneus brachii medialis, 8 N. cutaneus antebrachii, 9 N. thoracodorsalis, 10 Nn. subscapulares, 11 N. thoracicus longus, 12 N. subclavius, 13 N. suprascapularis, 14 N. dorsalis scapulae, 15 Art. axillaris, A Upper trunk, B Middle trunk, C Lower trunk, D Lateral cord, E Posterior cord, F Medial cord

ing arm is rotated slightly inward with the palm facing to the back. There are sensory deficits at the outer edge of the shoulder and the radial forearm.

The facultative C7 involvement is apparent from weakness of the triceps muscle and paresis in the pectoral muscle and finger flexors; the sensitivity disorder also includes the middle finger.

Lower plexus

Afferents from C8 and T1 proceed in the lower plexus. The sympathetic afferents flow from these segments to the head and the upper quadrants of the body, so that plexus lesions are often associated with an ipsilateral Horner syndrome (miosis, ptosis, enophthalmos, and quadrant-shaped sweat secretion disorder in the face, neck, arm, and hand as well as in the upper third of the thorax). There is motor

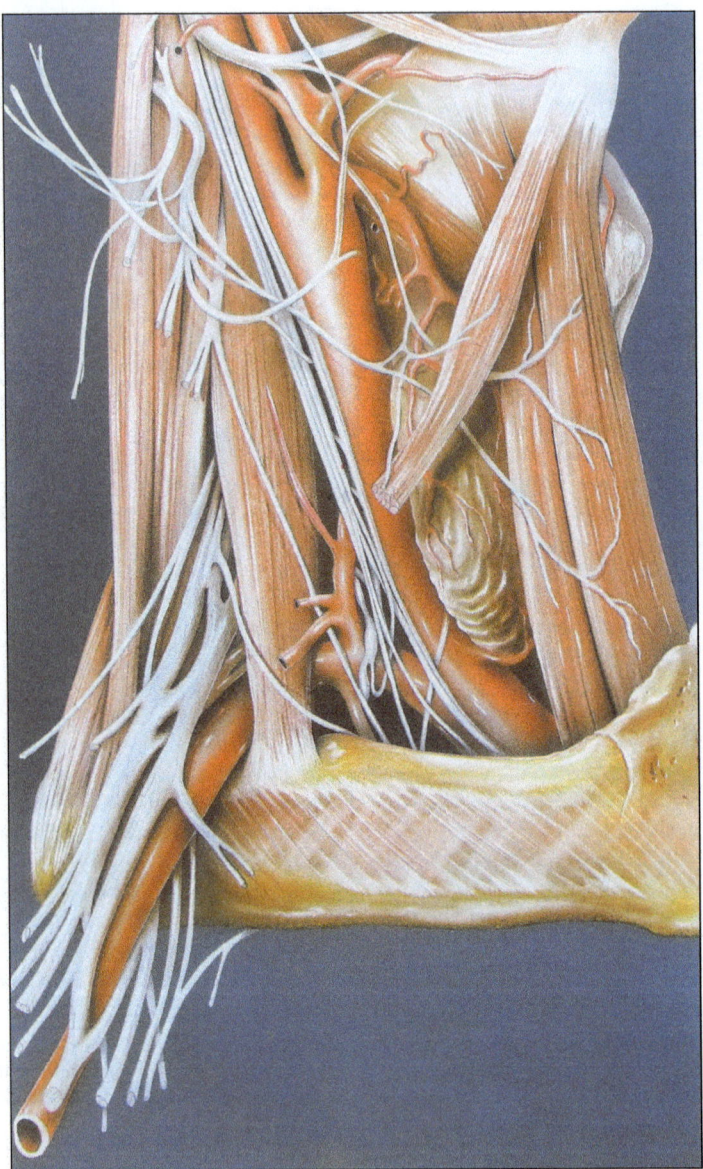

Figure 40 Anatomy of the brachial plexus in the neck with surrounding muscles and bones. The cervical plexus proceeds to the periphery between different neck muscles. The brachial plexus proceeds through the scalene hiatus. It is bordered in front by the anterior scalene muscle, which proceeds from the transverse processes of the 3rd to 6th cervical vertebrae to the first rib and in back by the medial scalene muscle, which also originates from the transverse processes, but proceeds behind the plexus to the floor of the hiatus to the first rib and inserts there

paresis of the small musculature of the hand, so that the tone of the healthy finger extensors extends the hand dorsally, forcing the fingers into a claw position. This picture is known as Klumpke paralysis. There is sensory involvement of the ulnar forearm and edge of the hand.

Mechanical injuries and their sequelae

Falls onto the shoulder cause considerable displacement in the shoulder due to forward, backward, or downward movement as well as from sideways and rotation movement of the head. Tension and stretching can cause tearing in connective tissue, nerve, and in vessels from capillaries up to large arteriovenous structures. In addition, there can be avulsion of nerve root fibers from the spinal cord. The consequences of such injuries are usually disastrous; the prospects of reinnervation are generally poor, root avulsions cannot be healed, and are often associated with the feared deafferentation pain. This poor prognosis makes caution in surgical interventions necessary. Reconstructive procedures are not simple. However, an experienced surgeon may consider surgery if there is a reasonable prospect for success, even if only minor. There is usually a chance of success for clean stab wounds in the plexus that can be treated while fresh (Figure 41).

Other injuries such as stretching or pulling or large areas of mechanical energy, such as those caused by gunshot wounds, lead not to circumscribed lesions, but rather to diffuse lesions in the whole plexus and are thus less suitable for surgery as the damage creates unfavorable conditions for a good prognosis. In an exploratory procedure, the plexus usually appears fully intact after an overextension trauma. The decisive factors are intrafascicular adhesions, scarring, strictures, neuromas, and loss of continuity that can affect a nerve fiber in more than one place. Nature therefore places insurmountable obstacles to reconstructive surgery. The quality of pain therapy, medical, and social rehabilitation determines the fate of a person with severe damage to the plexus.

The prognosis is significantly more favorable for damage caused by mainly demyelinating injuries and only to a lesser extent by tear damage such as loss of continuity due to extension and pressure. After traumas, a precise assessment of the muscula-ture of the extremity is often not possible due to pain and fractures. In this situation, EMG examinations can give important information for the prognosis such as denervation signs in the form of spontaneous activity. However, the examination is not helpful until 10–14 days after the accident, as the nerve will not be dead and degenerated enough to allow spontaneous muscle activity until that time.

Definite statements as to the prognosis are not possible in the acute stage of the injury; however, the concomitant conditions of damage allow useful assumptions to be made. If there is suspicion of root avulsion, myelography should be performed as soon as the patient's clinical condition allows in order for this diagnosis, which is so important for the prognosis, to be made.

Traction damage occurs most frequently in traffic accidents, especially with motorcycles or bikes, but also from athletic activities such as falls while skiing. Other lesions occur during shoulder luxation, during traumatic births, sternotomies, and thoracotomy, and as faulty positioning during surgery or other conditions with loss of consciousness. Plexus lesions may also arise in the sequel of a clavicle – or other bone fracture. Especially when during healing large callus formation are embedding and strangulating parts of the plexus. These mechanisms cause damage known as delayed paralysis. And not infrequently, damage arises from injuries from carrying heavy loads on one shoulder or from backpack straps pulling the shoulder downward. Stretching the C5–C7 root inflows can cause a lesion that looks like isolated damage to the long thoracic nerve and causes scapular alata. Axillary nerve damage can also appear as symptoms of such a mononeuropathy, although the damage is in the plexus. Innervation disorders that are often difficult to assess clinically can be discovered via changed action potentials of the motor units using EMG.

5.3.2 Damage from inflammation

Neuralgic shoulder amyotrophy is also known as plexus neuritis and Parsonage-Turner syndrome. It is a clinical diagnosis and there is no specific test that can confirm or refute the diagnosis with a sufficient degree of certainty. Taking the patient's pain history

5

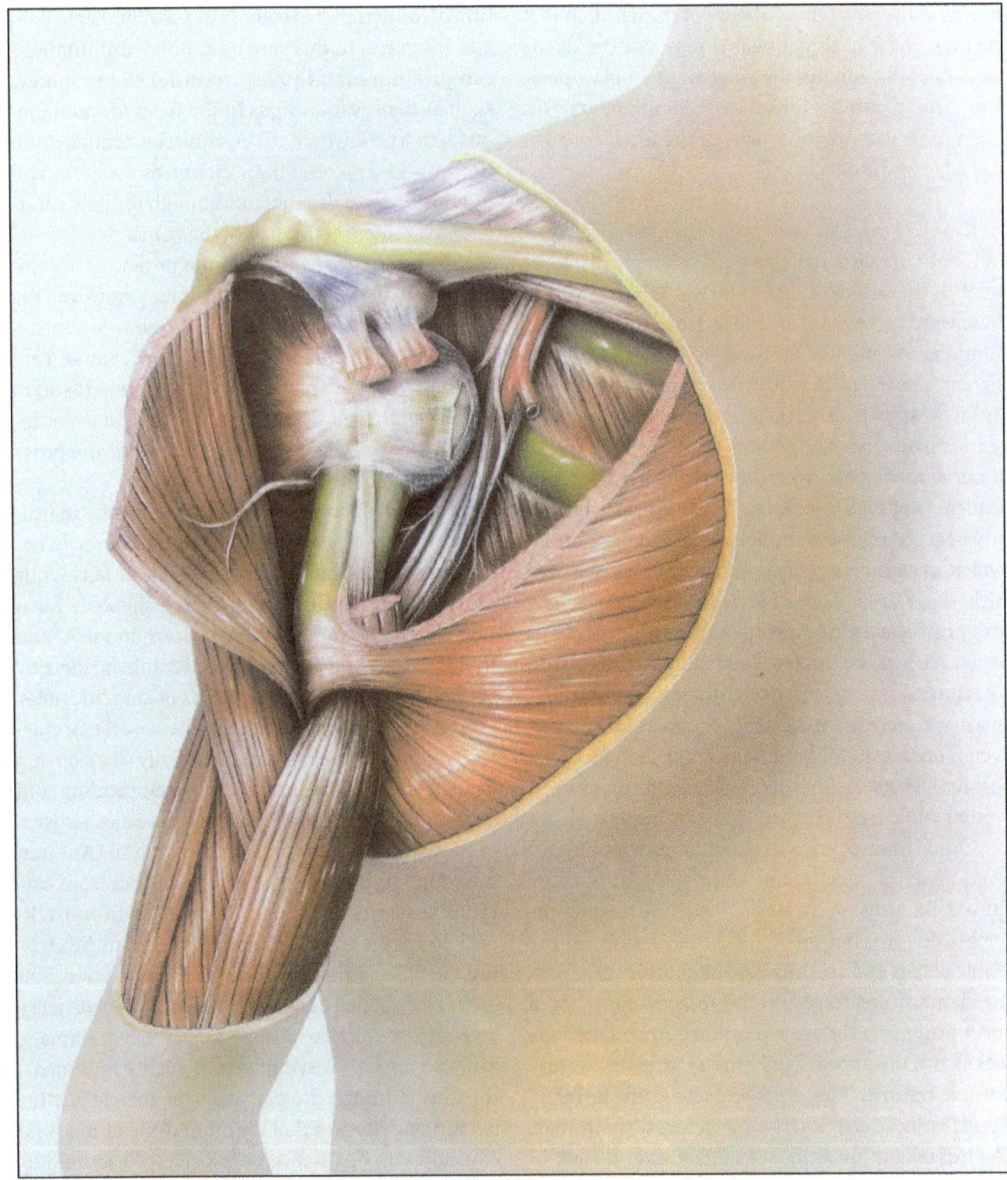

Figure 41 Shoulder luxation. The humerus head is luxated forward out of the shoulder joint and presses on the brachial plexus, the joint capsule is torn medially. Behind the brachial plexus, the axillary nerve proceeds to the deltoid muscle. The nerve is often damaged when the humerus is luxated backward or in fractures of the upper arm. Attempts to reposition with traction to the arm can easily cause traction trauma to the plexus. Not infrequently, shoulder luxation can occur during generalized epileptic seizures

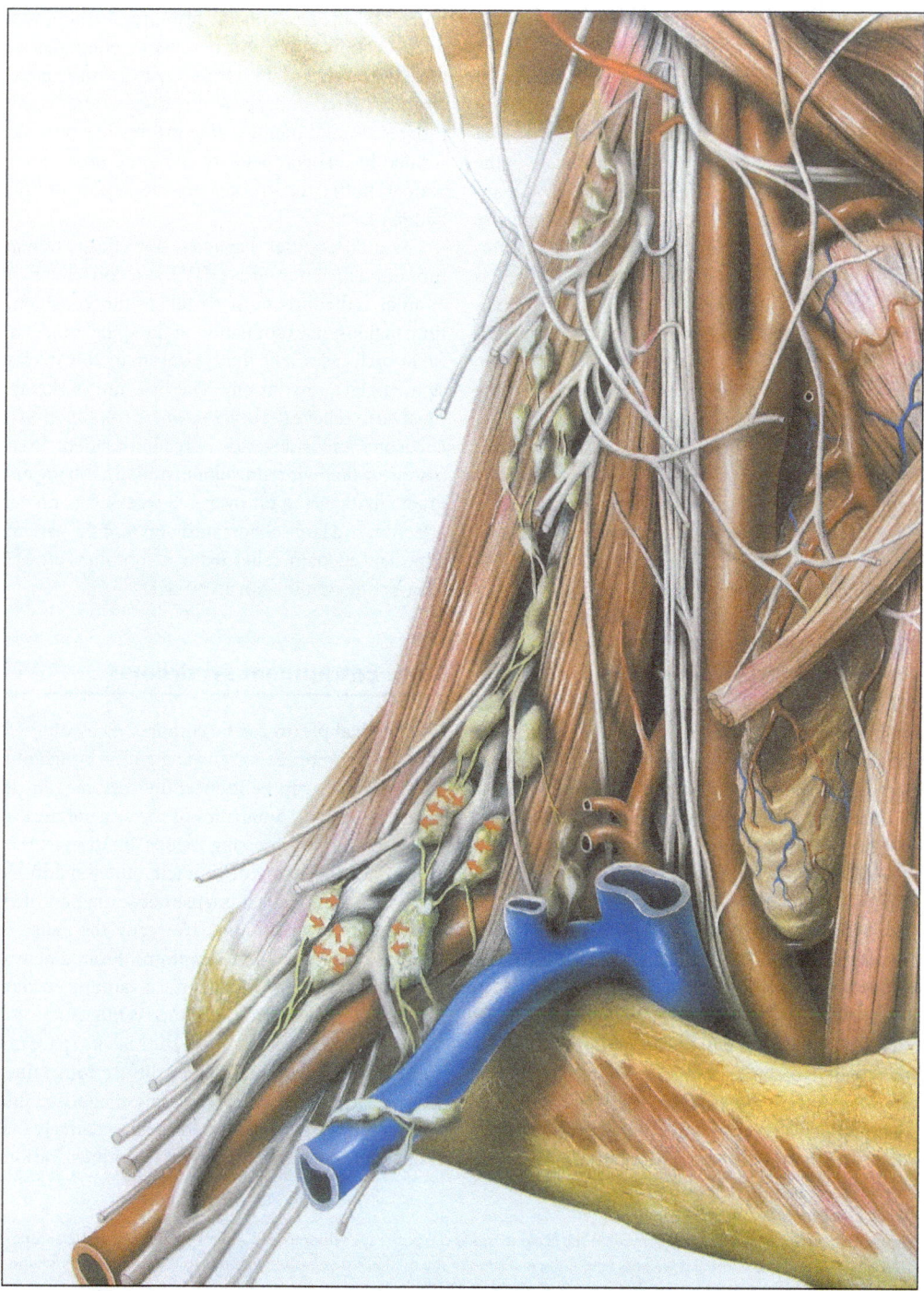

Figure 42 Neuralgic amyotrophy. Lymph node swelling can contribute to irritation of the plexus in neuralgic shoulder amyotrophy. The image also shows the close relationship of the lymph nodes to the cervical plexus. Diagnostic lymph node excision in this area often leads to lesions, especially of the accessory nerve, with paresis of the upper trapezius muscle

is important. A typical situation is a patient waking up in the morning with pain. The sharp pain is increasing and severe in the shoulder and upper arm radiating into the neck and hands. Limb paresis and occasionally sensory disturbances occur within hours or in the next few days. The pareses are in the domain of the long thoracic, suprascapular, and anterior interosseous nerves. Sensory symptoms are tingling in the superficial radial or lateral antebrachial cutaneous nerve regions. In some cases there is only a numb patch over the delta region in the axillary nerve distribution. It is puzzling that the pain is usually not in the same nerve territory distribution as the paresis, and both are often not in the same territory as the sensory symptoms. The pain is so intense that many patients are at first not aware of these additional symptoms. The often excruciating pain lasts for several weeks and does not respond to the usual analgesic treatment. Depending on the severity of the illness, it can take only a few weeks but even years until permanent healing is achieved.

The upper plexus segment is especially susceptible to this illness. Radicular or mononeuritic deficits do not contradict the diagnosis in any way (Figure 42), such as those of the median nerve (resembling anterior interosseous nerve syndrome) and radial nerve (resembling posterior interosseous nerve syndrome), or the lower brachial plexus with sympathetic nervous system involvement (resembling complex regional pain syndrome) and phrenic nerve involvement. The lumbar plexus can also be affected. The lumbosacral variant of neuralgic amyotrophy, also called "lumbosacral radiculoplexus neuropathy," typically occurs in patients with mild type 2 diabetes mellitus.

The cause of this illness is mostly unknown (see review Van Eijk et al. Muscle Nerve 2016;53:337). It is thought to be complex and multifactorial in some cases and in other cases hereditary. An interplay is assumed between environmental–mechanical factors with repetitive or strenuous motor activity and individual or genetic susceptibility. Autoimmune diseases, infections, and hereditary etiologies are most likely; an irritation from swollen lymph nodes was considered in the past. A special form of slowly progressive and painless amyotrophy has been described in diabetic patients and drug users. Some patients suffer recurrences in at least 25% of idiopathic cases.

As a differential diagnosis, borreliosis, human immunodeficiency virus (HIV), seroconversion, or another radiculitis must be taken into consideration, making an examination of the spinal fluid useful in such cases. The fluid is unremarkable in classic neuralgic amyotrophy. The first aim of therapy must be to relieve the severe pain. A combination of cortisone and analgesics is recommended. Prednisone or methylprednisolone, initially 100 mg and gradually tapering off over 2–3 weeks, has proven effective. In a large cohort study, 60 % of the patients experienced pain relief from a combination of a long-acting opioid with an NSAID.

5.3.3 Entrapment syndromes

The brachial plexus can be compressed by different structures. Compression and angulation by anatomical anomalies can be induced by cervical ribs, fibrous bands, and anomalies of the scalene muscle. Although the compressing lesions are diverse, they run under the name of thoracic outlet syndrome (Figure 40, 43). There is intense debate on how often these anatomical findings are really the cause of patients' complaints and symptoms. From a neurological point of view, this disorder is rare and encourages caution towards surgical interventions.

Entrapment syndromes in the scalene space can cause neurological, vascular, or diffuse complaints. The scalene syndrome is certainly diagnosed and operated too frequently, often causing damage that is out of proportion to the initial symptoms. Various

Figure 43 "Outlet syndrome." Entrapment in the scalene space can occur at its floor from a widened insertion of the anterior scalene (black arrows) or medial scalene, where the insertion of the medial scalene often extends to that of the anterior and forms the floor of the scalene space. Problems may arise when a cervical rib or an abnormal fibrous band (red arrow) associated with an unusually elongated transverse process of the C7 vertebra compresses the nerves. A vascular syndrome can develop when an abnormally high rib displaces the axillary artery upward. This can create an area of stenosis with the risk of post-stenotic dilatation

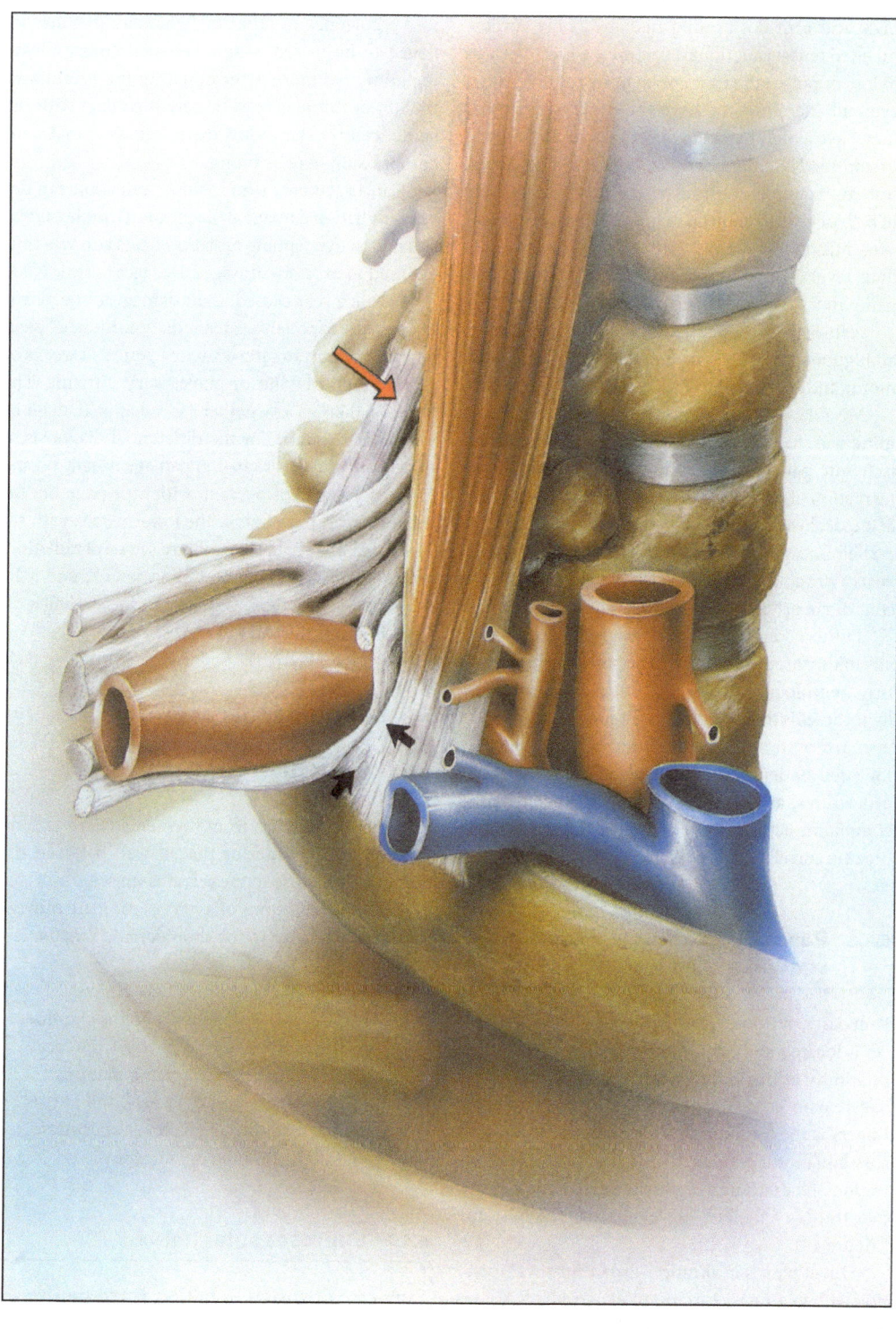

neck and arm positioning maneuvers are recommended to test functional strictures, which can lead to loss of pulse at the arm or stenosis sounds. However, all examinations have the disadvantage that they have a high percentage of positive results even in completely complaint-free persons. The Adson test, for example, compresses the subclavian artery in 80% of healthy subjects. These diagnostic maneuvers thus do not comply with the requirement for a high level of discrimination between normal and pathological findings.

Neurological deficits exist only if there are unambiguous symptoms from the lower plexus with motor and sensory deficits.

Vascular deficits manifest themselves in intermittent ischemia with fingers that turn white and ischemic pain. An occasional complication is the formation of post-stenotic thromboses that cause an acute occlusion of the subclavian artery or peripheral embolisms in the fingers. Usually, an x-ray reveals a pronounced cervical rib that compresses the subclavian artery from below.

Diffuse complaints are the most frequent, usually in the form of brachialgia or paresthesia with no firm neurological or vascular finding. One of many hypotheses is that there is an irritation of the subclavian artery in the costoclavicular space. Especially for such patients, caution is advised for surgical procedures, as it is by no means certain that the complaints actually have a surgically approachable organic correlate.

5.3.4 Pancoast syndrome, radiation sequelae

Pancoast syndrome is frequent when bronchial cancer is located in the tip of the lung. Due to the close proximity of the lower plexus, it can be infiltrated. Severe pain in the forearm including the two ulnar fingers is the leading symptom, later there are sensory and motor deficits. Due to the additional infiltration of the stellate ganglion located in the vicinity, two thirds of patients develop Horner syndrome (Figure 44).

Direct plexus infiltration also occurs in association with mammary carcinoma as a result of osseous and lymph metastases. It can also appear as a local-ized recurrence of other malignancies that are believed to be in remission. The symptoms, always including pain, are determined by the location of the tumor in the plexus. If there is contact with the spine, central symptoms expressed as spinal cord compression may be found.

Cancer patients who required radiation can develop radiation damage at the plexus if the plexus was situated in overlapping radiation fields and was thus exposed to excessive dosages. Dosages of over 6,000 rad within a year caused tissue damage to the plexus in over 70% of patients. Such radiation-induced plexopathies can occur after a latency period of weeks or many years, making diagnosis very difficult. The question then arises whether it is radiation damage or a recurrent tumor. For the differential diagnosis, it may be helpful that radiation damage usually occurs in the upper plexus segment with little pain, but tumors occur more often in the lower plexus with severe pain. Skin changes are often a sequel of radiation, lymph edema occurs in both illnesses. CT and MRI examinations are often very helpful for making a diagnosis in this difficult- to -access region.

5.4 Shoulder nerves

5.4.1 Long thoracic nerve

The nerve originates from segments C5 to C7 and transverses the anterior thorax wall between the plexus and clavicle to the serratus muscle.
- Clinical symptoms of a nerve lesion are movement problems in the shoulder and scapula alata.
- Damage can occur to the nerve from:
  - Tension and compression on the shoulder girdle from: backpack, sports, work, accident, or intraoperative manipulations
  - Surgical interventions at neck and plexus such as from placing a plexus anesthesia
  - Neuralgic shoulder amyotrophy

5.4.2 Suprascapular nerve

The nerve originates from C5/C6 and transverses the supraclavicular foramen of the shoulder blade,

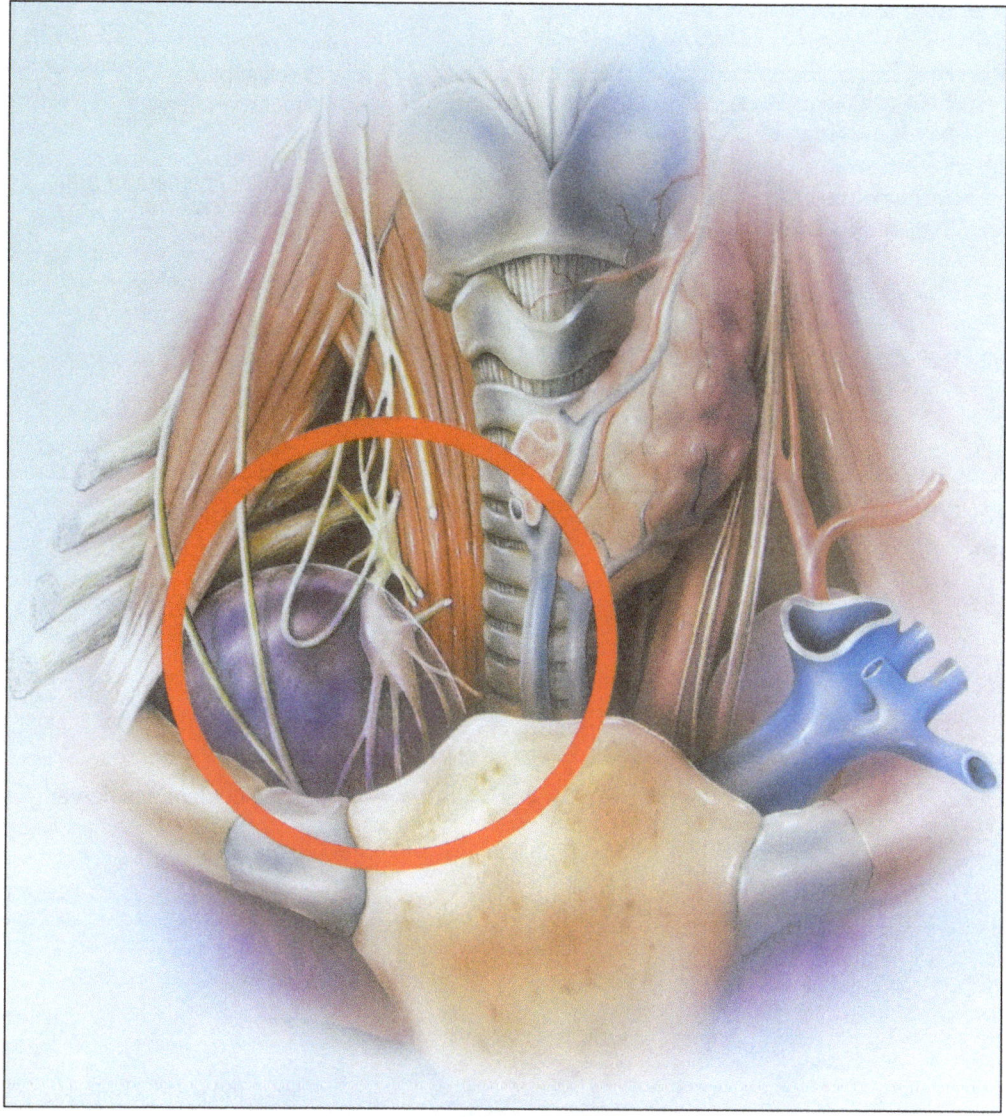

Figure 44 Anatomy of the upper thoracic aperture and the scalenovertebral triangle. The pleural dome is located in close proximity to the sympathetic trunk and the stellate ganglion as well as to the lower segment of the brachial plexus

supplies the supraspinatus muscle and shoulder joint segments, then transverses the spina scapulae and innervates the infraspinatus muscle.

— Clinical symptoms of lesions are often shoulder pain, atrophy of the muscles supplied, and limited shoulder abduction.

— Damage to the nerve can be caused by: Compression at the supraclavicular foramen Scapula fractures Neuralgic shoulder amyotrophy

— Differential diagnosis: Rupture of the rotator cuff

5

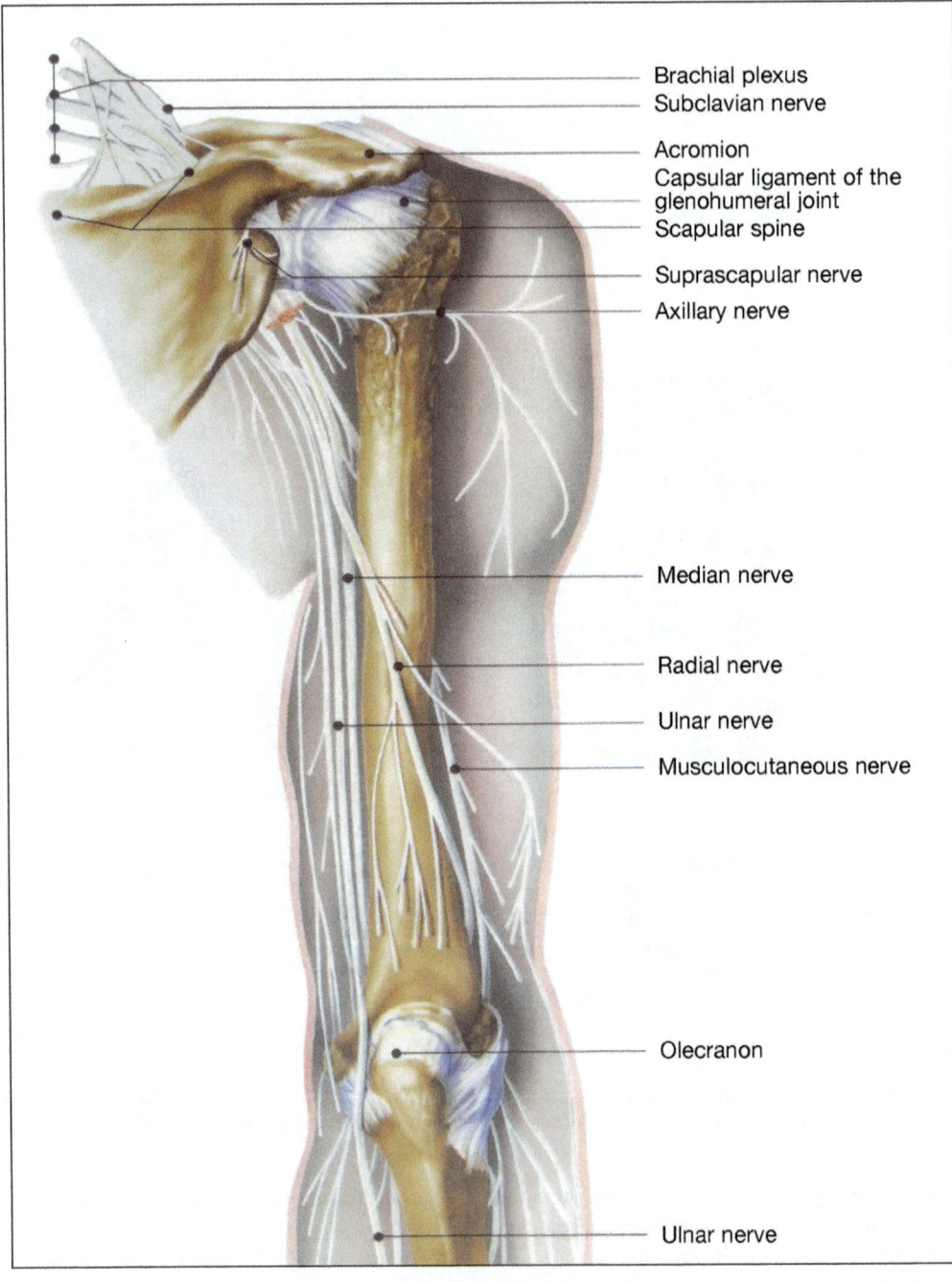

Brachial plexus
Subclavian nerve

Acromion
Capsular ligament of the
glenohumeral joint
Scapular spine

Suprascapular nerve
Axillary nerve

Median nerve

Radial nerve

Ulnar nerve

Musculocutaneous nerve

Olecranon

Ulnar nerve

Figure 45 Nerves in shoulder and arm in relation to osseous formations

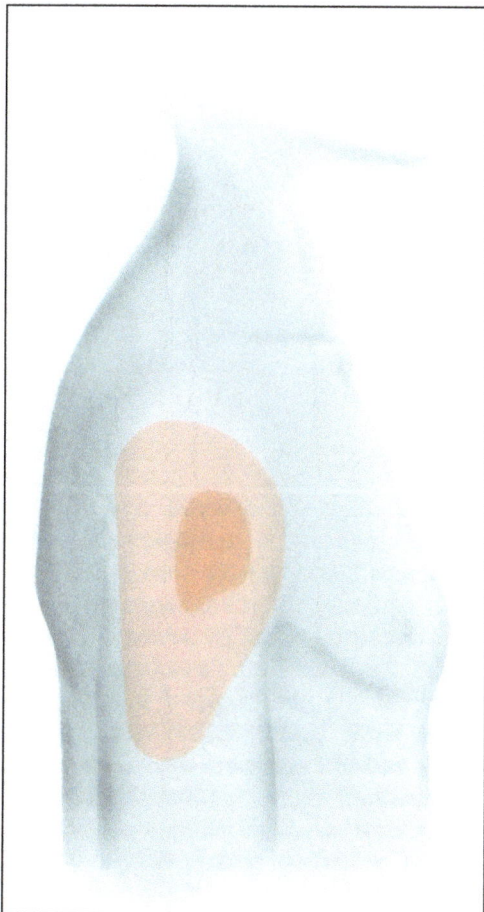

Figure 46 Sensory skin supply of the axillary nerve with intermediate (light red) and autonomous (red) innervation zone

5.5   Axillary nerve

The nerve originates from C5/C6, proceeds through the posterior plexus fascicle, exits through the axillary space, proceeds around the humerus head, and innervates the deltoid muscle, teres minor muscle, and the skin over the shoulder (Figures 45, 46).
- Clinical symptoms of an axillary nerve lesion are abduction paresis with atrophy of the deltoid muscle, occasionally compensated by the supraspinatus muscle.
- Damage to the nerve is caused by: injuries to soft tissue, bone fractures, and lesions of the plexus with symptoms of mononeuropathy.

5.6   Musculocutaneous nerve

The nerve originates at C5–C7, proceeds in the lateral plexus fascicle, and innervates the brachial biceps muscle and brachialis muscle. Its sensory branch, the lateral antebrachial cutaneous nerve, supplies the skin of the radial forearm up to the wrist.
- Clinical symptoms of a nerve lesion are, depending on the location of the lesion, a combined or isolated deficit pattern of biceps weakness and deficits in the sensory supply territory.
- Damage to the nerve can be caused by:
  - Fractures and luxations of the shoulder
  - Neuralgic shoulder amyotrophy
- Differential diagnosis:
  - C6 syndrome
  - Rupture of the biceps tendon

5.7   Radial nerve

The nerve originates from C5 to T1 and the posterior fascicle proceeds dorsally of the axillary artery in the axilla, and after supplying the triceps muscle continues with the circumflex humerus artery in the radial groove around the humerus shaft to the elbow. It supplies the long and short brachioradial muscle and the extensor carpi radialis muscles. It then branches into the purely sensory superficial ramus and innervates the skin of the radial forearm and hand. As interosseous posterior nerve or ramus profundus, it continues then as a purely motor nerve through a canal in the supinator muscle and innervates all hand and finger extensors (Figure 47).
- Clinical symptoms, depending on the level of the lesion, as indicated in Figure 48 by numbered arrows, are various types of sensory or motor paresis.
  1. A lesion at this level causes sensory and motor paresis of all areas supplied by the radial nerve, the triceps reflex is thus also absent.
  2. Triceps muscle paresis usually occurs with a fracture of the humerus shaft; park bench compression lesions are located distal to this level. They usually affect the nerve in its course in the sulcus after the triceps muscle has branched off. Clinically, there is thus usually no paresis of the triceps muscle.

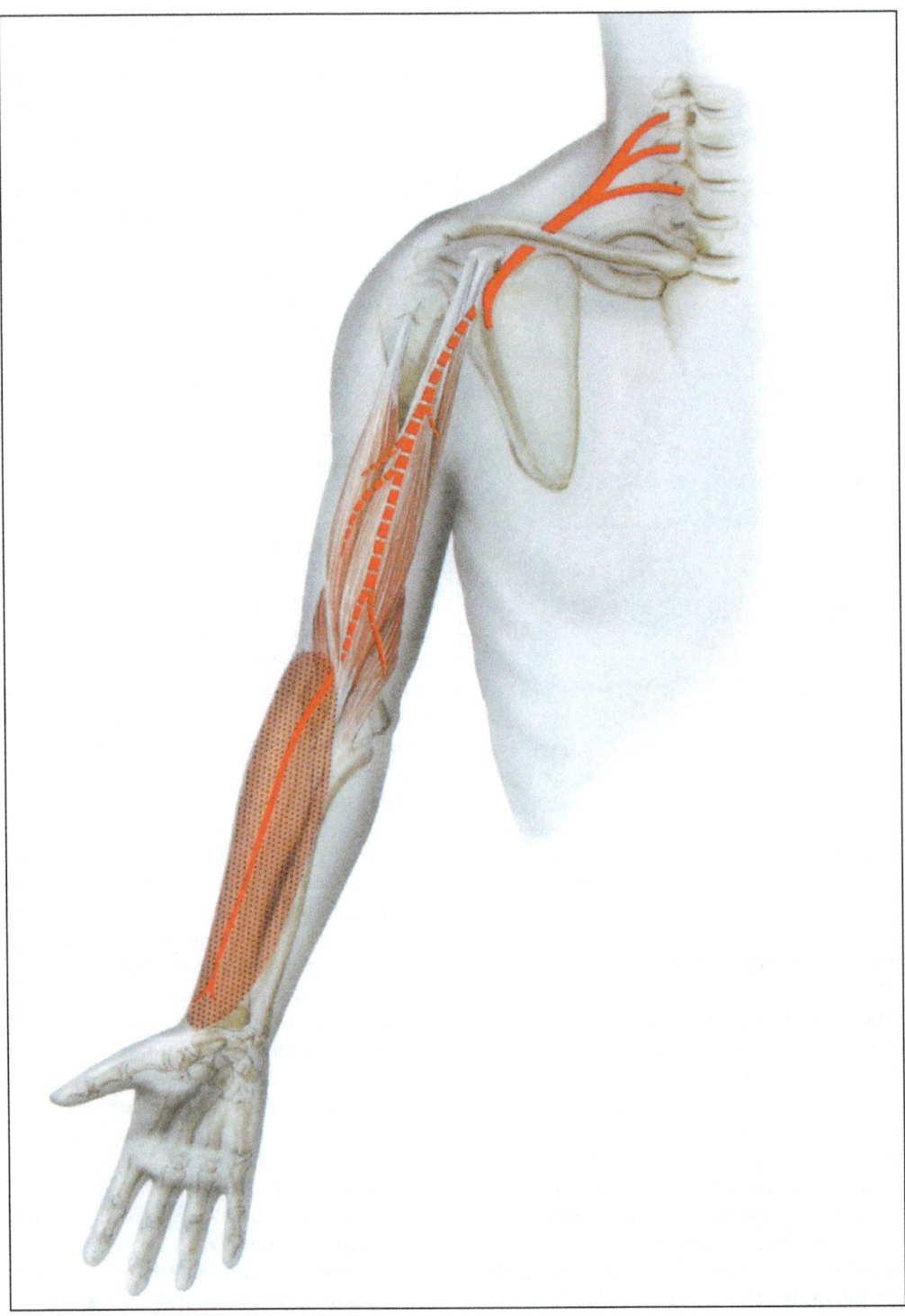

Figure 47 Anatomy of the musculocutaneous nerve

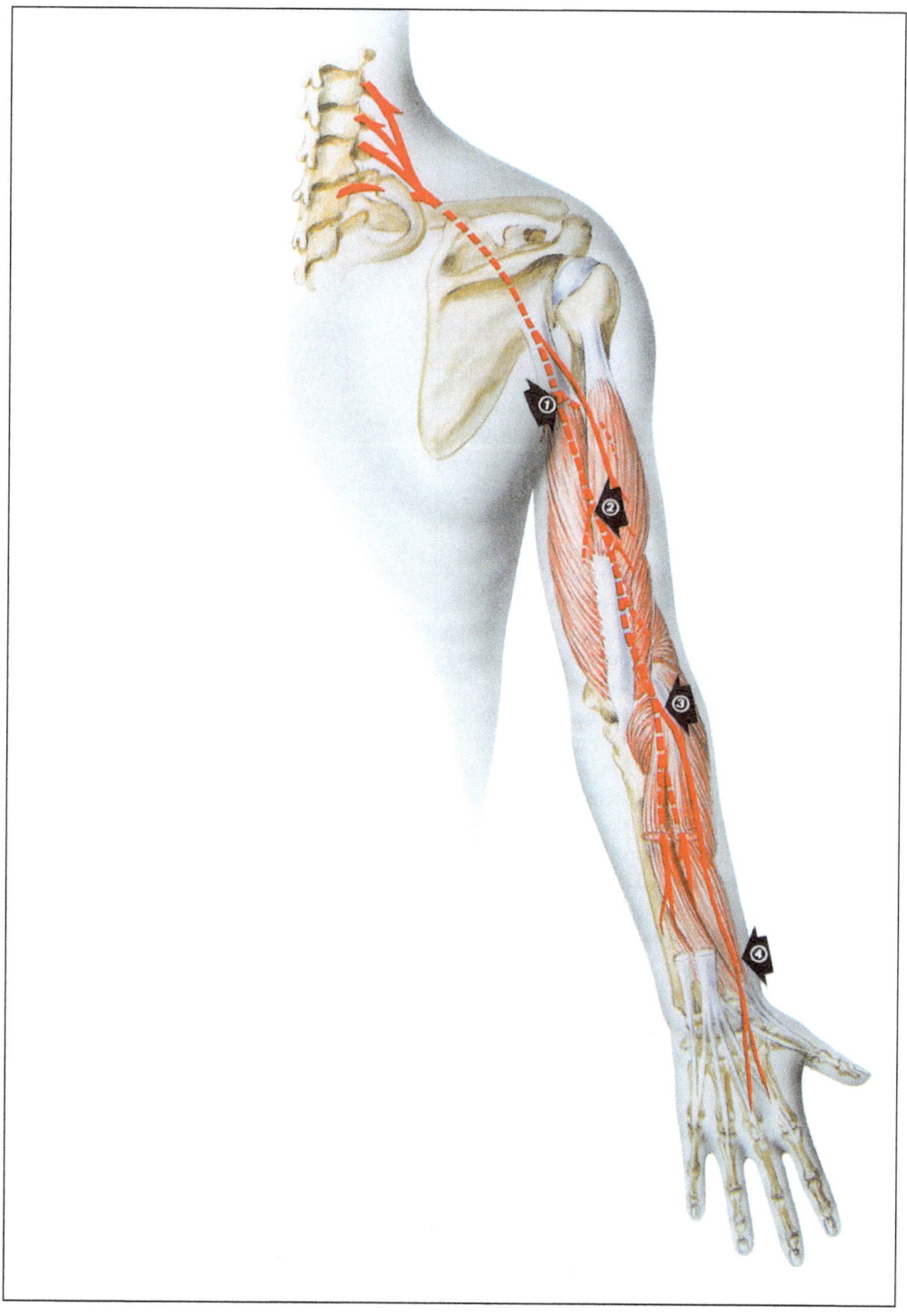

Figure 48 Frequent damage sites on the radial nerve. Numbers 1 to 4 are explained in the text

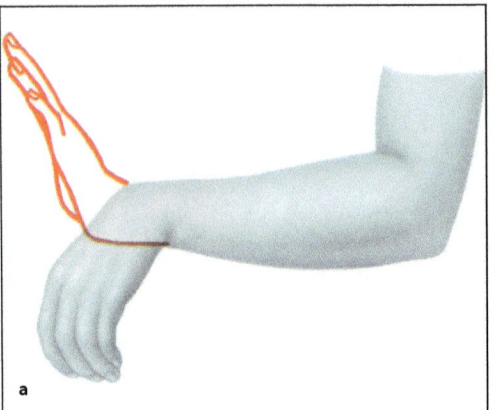

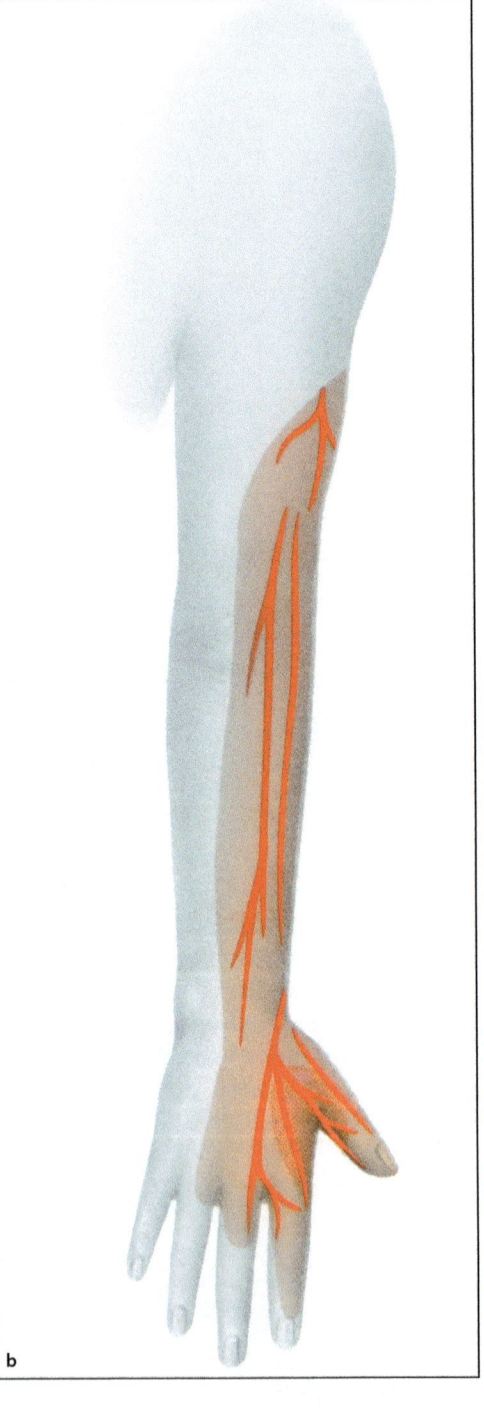

Figure 48a, b **a** Damage to the main stem in the upper arm leads to a deficit of the extensor musculature – the socalled drop hand. It is characteristic of radial nerve paralysis. It makes it impossible to extend the wrist or metacarpophalangeal joints. **b** Autonomous (dark red) and intermediate (light red) sensory innervation area of the radial nerve

3. Lesions above the supinator canal. They often cause a drop hand (Figure 48a) due to paresis of the brachioradial muscle and the extensor carpi radialis muscle, also sensory deficits on the skin, but sometimes only pain. The supinator syndrome, nerve damage in the canal, has a purely motor deficit pattern. As the lesion affects only the radial extensors of the finger extensors and not those of the arm, there is no drop hand.
4. Compression or strictures at the wrist cause only sensory deficits.

— Damage of the nerve in Figure 48b is caused by:
1. Pressure from crutches in the armpits
2. Upper arm shaft fracture or positioning damage (park bench, operating table, edge of the bed)
3. Supinator syndrome (entrapment, occupational paresis)
4. Bracelet, handcuffs, dialysis shunt
5. Compression lesion in fingers, e.g., from using scissors

— Differential diagnosis:
— C6, C7 syndrome
— Rupture of the tendon of the extensor pollicis longus muscle in a radius fracture
— Mononeuritis
— Volkmann's contracture of the flexors

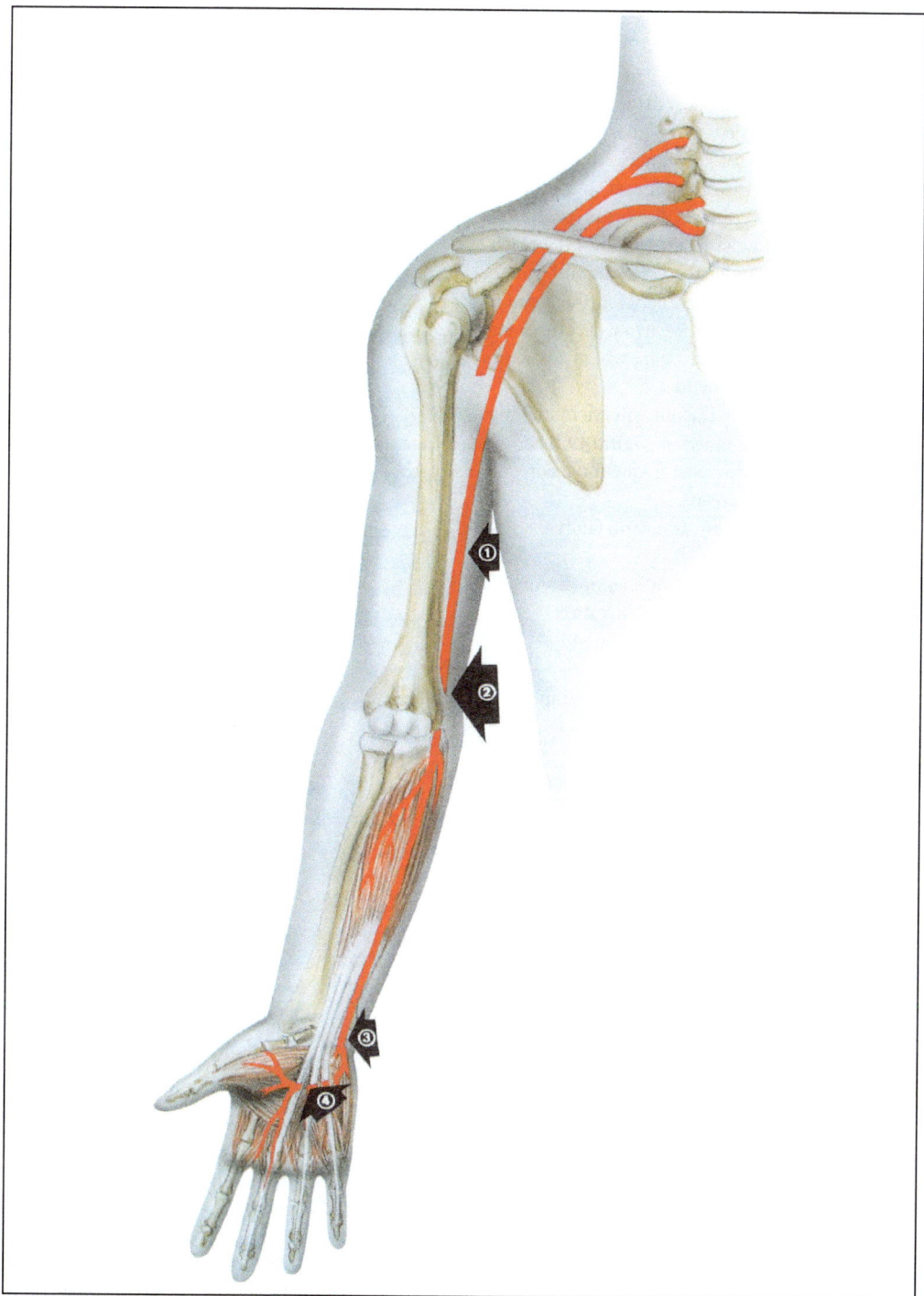

Figure 49 Course of the ulnar nerve and its damage sites. 1 Honeymoon palsy, 2 Lesions in the sulcus, 3 Loge de Guyon, 4 Biker's lesion

5.8   Ulnar nerve

The nerve originates from C8/T1 and the medial fascicle, follows the brachial artery in the bicipital groove of the upper arm, proceeds in the ulnar groove around the medial epicondyle of the humerus to the forearm flexor side, supplies the ulnar flexors, and proceeds protected under the flexor digitorum profundus muscle to the ulnar wrist. There (Figure 49) it reaches the carpal bone under the volar carpal ligament and a fibrous band stretched between the pisiform and hamate bones, the loge de Guyon, over the transverse carpal ligament to the palm, where it supplies the small muscles of the hand.

— Damage sites where the ulnar nerve frequently suffers lesions are shown in Figure 49. The damage site and the most frequent type of damage caused there are:
1. Upper arm: compression and faulty positioning
2. In the ulnar nerve groove: compression and connective tissue degeneration
3. In the loge de Guyon: entrapment syndromes
4. In the palm of the hand: direct compression damage

— Clinical symptoms of an ulnar lesion:
Sensory deficits (Figure 50) at the ulnar half of the ring finger and entire little finger. Deficits that affect only the back of the hand and the dorsal side of the little finger and half of the ring finger stem from dorsal ramus lesions. This purely sensory branch of the ulnar nerve branches off from the main branch of the nerve in the middle of the forearm. Deficits that affect only the palmar side of the fingers stem from palmar ramus lesions. This purely sensory branch in the wrist originates from various locations.

In lesions in the sulcus and above the sulcus, motor deficits affect the ulnar flexors and the small hand muscles. Isolated hand muscle paresis indicates that the lesion is further distal in the forearm. If the lesion is even further distal, there is no sensory damage, which is typical for a purely motor lesion of the ramus profundus in the loge de Guyon region. Damage to the ulnar nerve leads to the formation of the so-called claw hand (Figure 51).

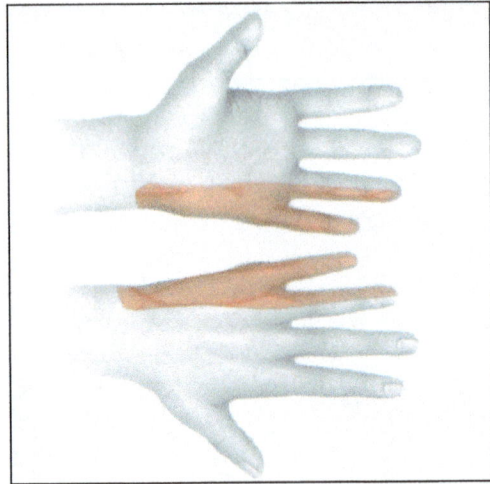

Figure 50 Sensory skin innervation of the ulnar nerve

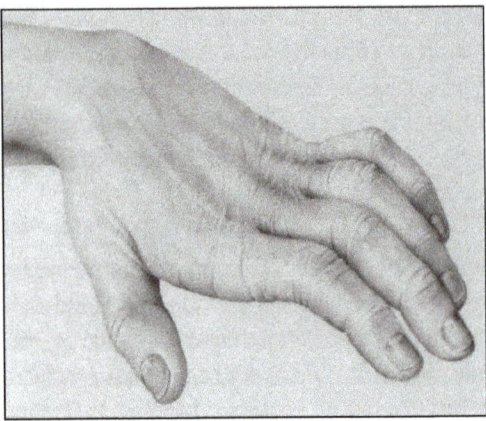

Figure 51 Picture of the claw hand

The configuration depends on the innervation pattern of the intrinsic hand muscles. A three-finger claw hand results when the lumbrical to the long finger is innervated by the ulnar nerve. Complete high ulnar nerve lesions produce a clawed hand when the intact median nerve supplies the flexor digitorum profundus to the ring finger and the little finger.

The fingers in the metacarpophalangeal joints are overextended, but flexed in the interphalangeal joints. This characteristic finger position is triggered by paralysis of the interosseous muscles that are responsible for extension of the distal phalanges. Deficits in the flexors result in the finger position shown in Figure 51 due to the predominance of the extensors. As the small finger and the adductors of the thumb are paralyzed, the thumb and small finger can no longer meet to form an "O." Rapid extension movements, such as when flicking away a ball of paper with the thumb, are no longer possible (Figure 52). The spaces between the interphalangeal joints are sunken due to atrophy of the interosseous muscles.

A deficit of sweat gland activity in the sensory innervation territory of the ulnar nerve in the ninhydrin test is indicative of a peripheral lesion; an unremarkable sweat test indicates a radicular lesion (Figure 53).

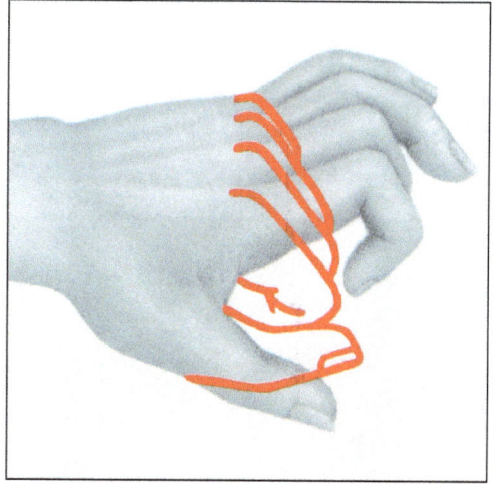

Figure 52 The flick sign

5.8.1 Ulnar groove and loge de Guyon

The sites where the ulnar nerve is most frequently damaged should be described in more detail. They are the ulnar nerve groove at the elbow and the loge de Guyon at the hand. In both regions, generally mechanical irritations lead to damage and anatomical anomalies that can further complicate the clinical picture.

In the sulcus ulnaris (Figure 54), the ulnar nerve groove, the nerve is very exposed on the extensor side of the elbow joint and is often subjected to habitual, especially mechanical, and thus demyelinating lesions. The nerve is often attached to the bone in mechanically irritated, therefore proliferating connective tissue and constricted. Other possible lesions are delayed paralysis after fractures and callus formation, elbow luxation, degenerative osseous changes, a bone spur above the groove, and direct stab, cut, and compression traumas.

Entrapment can occur above the groove from the intermuscular septum to the triceps muscle in the arcade of Struthers.

Below the groove, the tendon arcade between the two heads of the flexor carpi ulnaris muscle or aponeurosis of segments of the triceps muscle can cause a cubital tunnel syndrome. The individual lesion sites cannot be distinguished clinically. Therefore, in surgical neurolysis of the nerve, all regions must be carefully inspected and corrected.

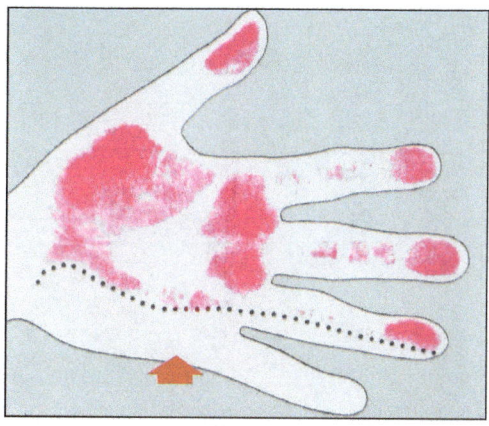

Figure 53 Ulnar lesion with denervated sweat glands leads to this typical deficit pattern in the ninhydrin test

Incomplete decompression of the nerve is the main cause of therapy failures. Different methods are used for neurolysis of the nerve. Mobilization from the sulcus is not unproblematic, as new irritations can occur if mobilization is inadequate. If the segment mobilized is too long, vascular supply disorders and ischemic damage can occur. Generally, the nerve is therefore left in its bed and only external neurolysis is performed. Interfascicular neurolysis is no longer carried out.

In the loge de Guyon (Figure 59), the palmar ramus branches out with several branches to supply

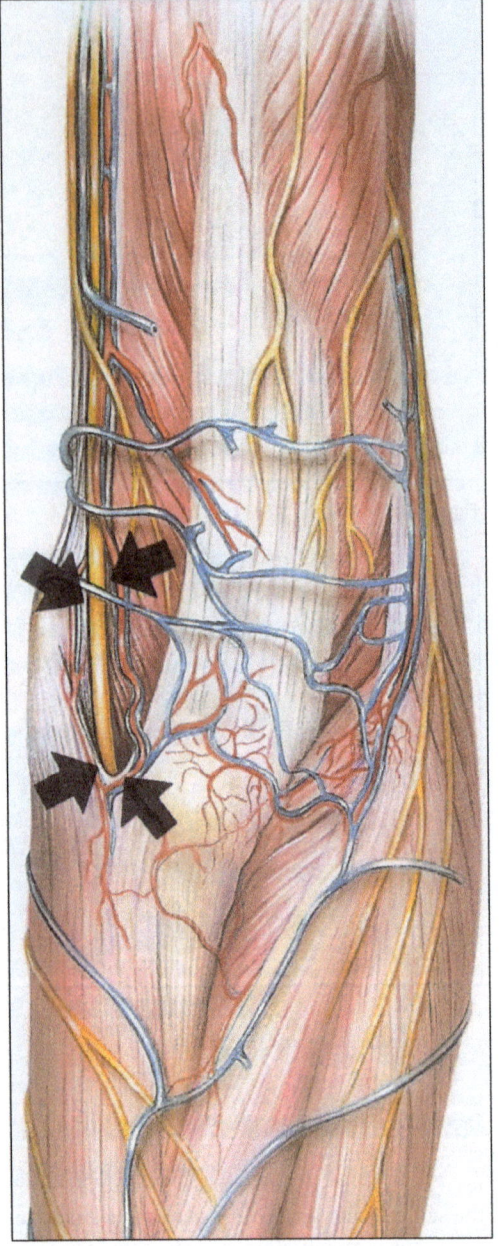

Figure 54 Anatomical situation in the area of the ulnar nerve groove

the small hand muscles. Therefore, lesions at various sites in the loge can affect individual muscles in various ways. The precise localization of the damage site can be left to fine neurological diagnostics. It is more important that in all cases, the patient complains of lack of dexterity of the hand – the more pronounced the loss of dexterity, the higher the demands made on fine motor skills. Since the lesion affects only the motor segment of the peripheral nerve, no disorder in the sweat test is to be expected.

The following anomalies must be considered as the cause of damage to the nerve: ganglia, bone fractures, and luxations, as well as aberrant muscles and vascular changes of the ulnar artery. Damage is most frequently caused by hyperextension in the wrist during cycling or from direct pressure in the palm of the hand, for example, from using a screwdriver.

5.9   Median nerve

The nerve originates from C6–T1, its fibers proceed in the medial and lateral fascicle and unite to form the median nerve loop. In the biceps groove, the nerve continues to the elbow joint, where it abuts the brachial artery medially and the biceps tendon laterally. Between the heads of the pronator teres muscle, the median nerve sinks into the forearm musculature, first innervating the pronator, the flexor digitorum superficialis muscle, and the flexor pollicis longus muscle. It later gives off the sensorimotor anterior interosseous nerve, continues to the carpal tunnel, and there innervates the small hand muscles. The interosseous nerve supplies sensory innervation to the joint capsule and the periosteum at the wrist, motor innervation to the flexor digitorum profundus, flexor pollicis longus, and pronator quadratus muscles. Figure 55 shows sensory skin innervation.

— Damage sites where the median nerve frequently suffers lesions are shown in Figure 56. The damage site and most frequent types of damage are:
1. At the upper arm, from compression and positioning during anesthesia or ischemia
2. At the forearm from misplaced injections or entrapment lesions in the area of the

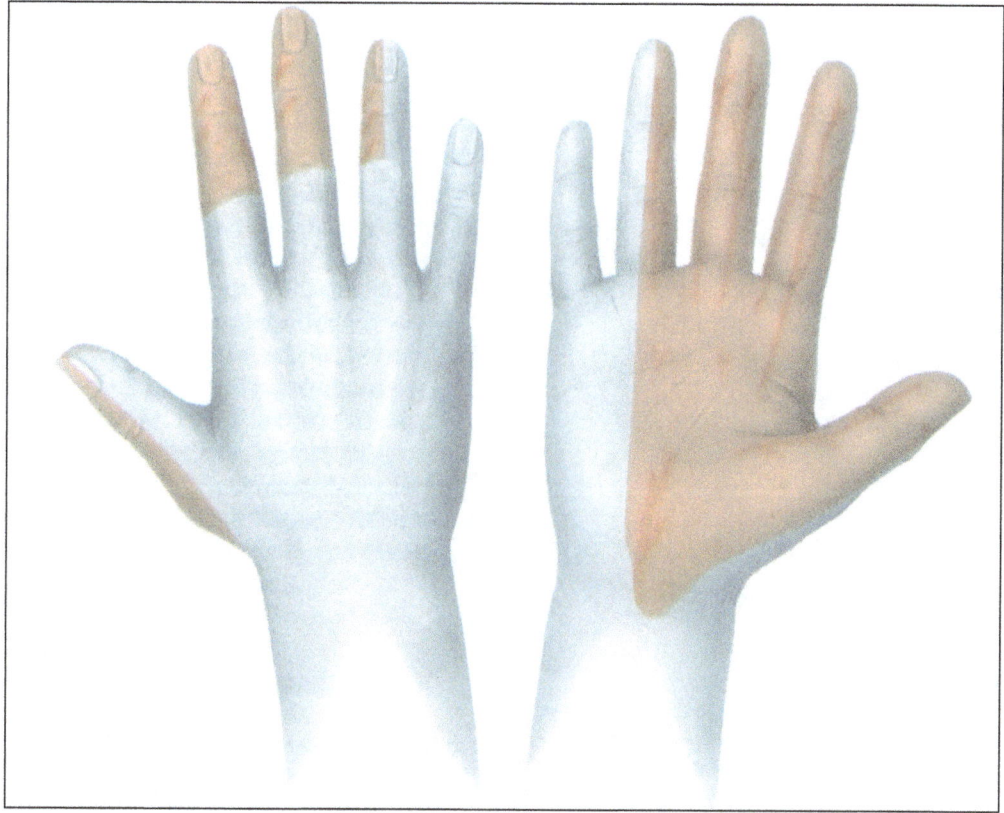

Figure 55 Area of sensory skin innervation by the median nerve

pronator teres muscle and interosseous nerve

3. At the front of the wrist, from suicidal or accidental cut or stab wounds

4. At the wrist, from entrapment lesions in the carpal tunnel

▬ Clinical symptom of a typical median nerve lesion when making a fist is the preacher's hand, as the finger flexors are paralyzed and the fingers remain extended (Figure 57). The bottle test (Figure 58) also helps to display muscle paralysis.

At the forearm, the median nerve can suffer pronator teres syndrome or damage to its interosseous nerve. Both of these are rare lesions and are not easy to differentiate. To make things even more difficult, the only clinical symptom – diffuse pain that radiates into the forearm, the palm, or the wrist – is common to both entrapment syndromes.

In the pronator teres syndrome, the pronator teres muscle is frequently tender or Tinel's sign can be elicited by tapping. Muscle paresis is never very pronounced, the muscles innervated by the interosseous nerve may even be unaffected.

The interosseous nerve can be irritated by strands of connective tissue below the pronator teres muscle (Kiloh-Nevin syndrome). The clinical feature of this syndrome is paralysis of the flexor pollicis longus, pronator quadratus, and a variable amount of the radial part of the flexor profundus serving the index finger. The "pinch sign" or circle test is useful for diagnosing the lesion. The patient cannot press the tips of the thumb and index finger together to form an "O," instead, only the pads of the fingers meet. Paralysis in the flexor digitorum

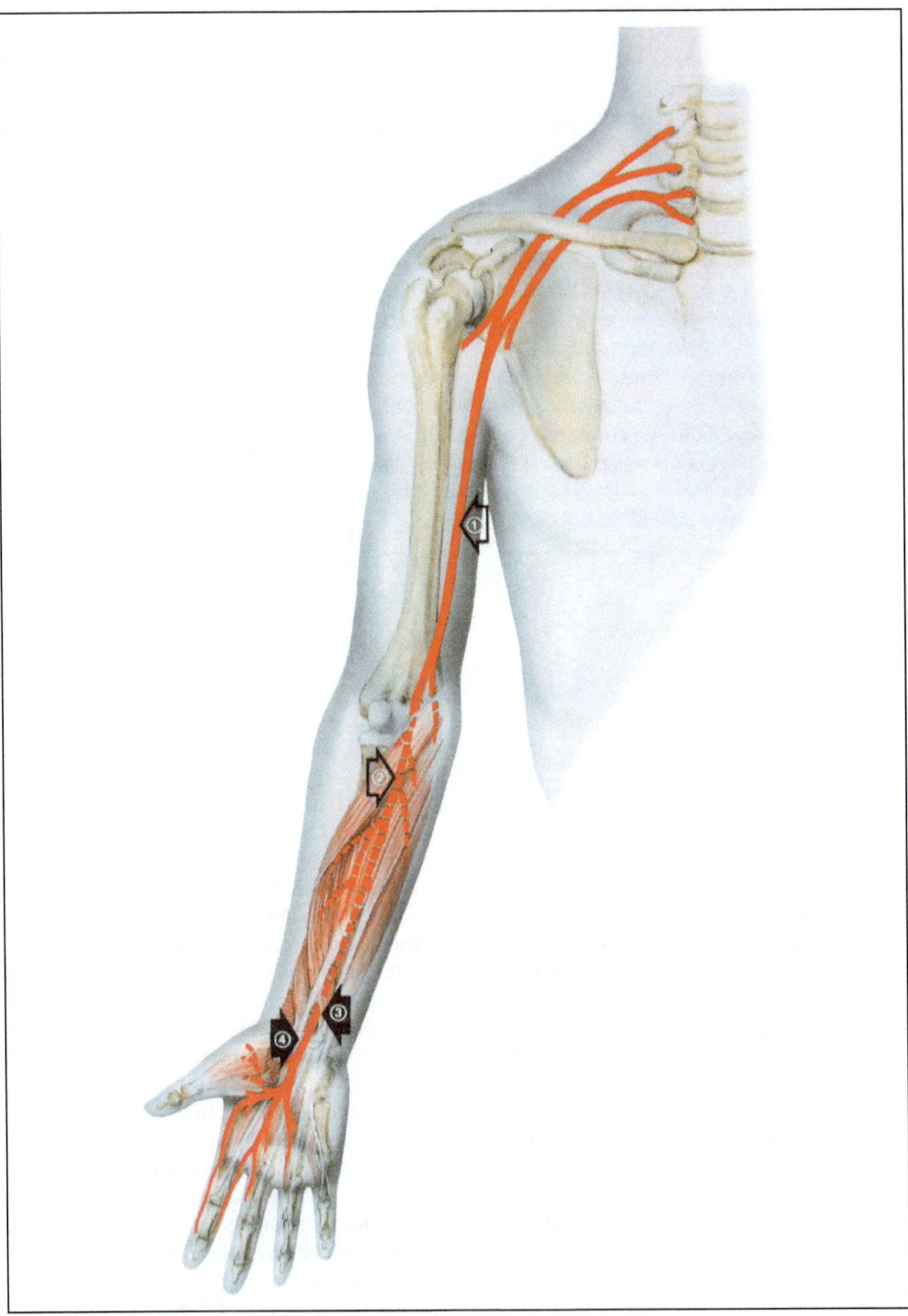

Figure 56 Course of the median nerve and frequent damage sites

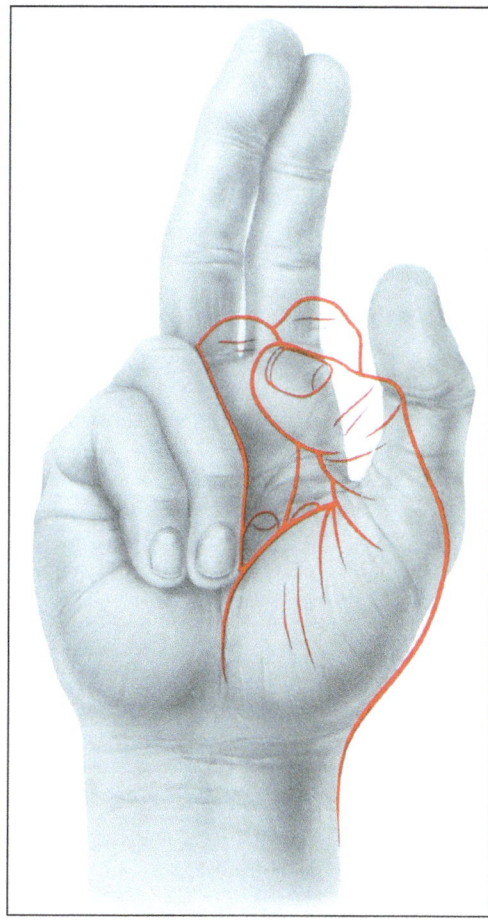

Figure 57 "Preacher's hand" in median nerve lesion. The weakness of thumb abduction is seen in the inability to grasp a bottle properly. If the web between thumb and index finger does not fit the curvature of the bottle, the bottle test is positive (Figure 58)

profundus muscle of the index finger does not allow flexion, just as paralysis of the flexor pollicis longus muscle does not permit the thumb to be flexed. However, lesions of the interosseous nerve and muscle paralysis often recede spontaneously.

5.9.1 Carpal tunnel syndrome

The carpal tunnel syndrome is by far the most frequent entrapment syndrome of peripheral nerves (Figure 59). Women are affected almost twice as of-

ten as men; usually the working hand is impaired. Maximum manifestation lies between ages 45 and 65. The carpal tunnel is formed by the arch of carpal bones spanned by the transverse carpal ligament. Toward the hand, the tunnel narrows like a funnel. In addition to the median nerve, nine muscle tendons and their sheaths share this space. Any space-occupying process first affects the soft, thus weakest, part of the median nerve.

Pathogenesis

Fine vessels run through the median nerve, as in every other nerve. Any pressure on the nerve leads to circulation disorders with ischemia that can trigger sensory irritations such as paresthesia. Hand and thus tendon movement promotes circulation in the carpal tunnel and improves the symptoms. But pressure for a longer period causes interand intrafascicular edema as well as anoxia. The edema covers a long segment and causes pseudoneuromatous swelling of the median nerve before the carpal tunnel. From this point on, two damage patterns compete against each other. Anoxia and compression cause demyelination. On the other hand, the edema attracts fibroblasts that proliferate and cause severe mechanical damage to the nerve by peri- and endoneural scarring. The ratio of these reversible and irreversible changes determines the extent to which the nerve can recover after decompression and neurolysis.

Clinical symptoms

The initial clinical symptoms are only pain and paresthesia, which, because they occur mainly at night, have given the syndrome the name "brachialgia paresthetica nocturna." Symptoms occur not only in individual fingers innervated by the median nerve, but are also found proximal to it as referred pain in the elbow, the lateral shoulder, and even up to the neck. The pain wakes the patients at night and they typically report that moving their fingers brings rapid and clear relief. Objectifiable deficits do not occur in this initial phase. The next stage of damage is sensory deficits that quickly lead from discrete hypoesthesia and hypoalgia to such severe disorders of sensory afferents that fine motor activity is no longer possible without visual control (knitting, unbuttoning shirt). Parallel to this, rarely isolated, is

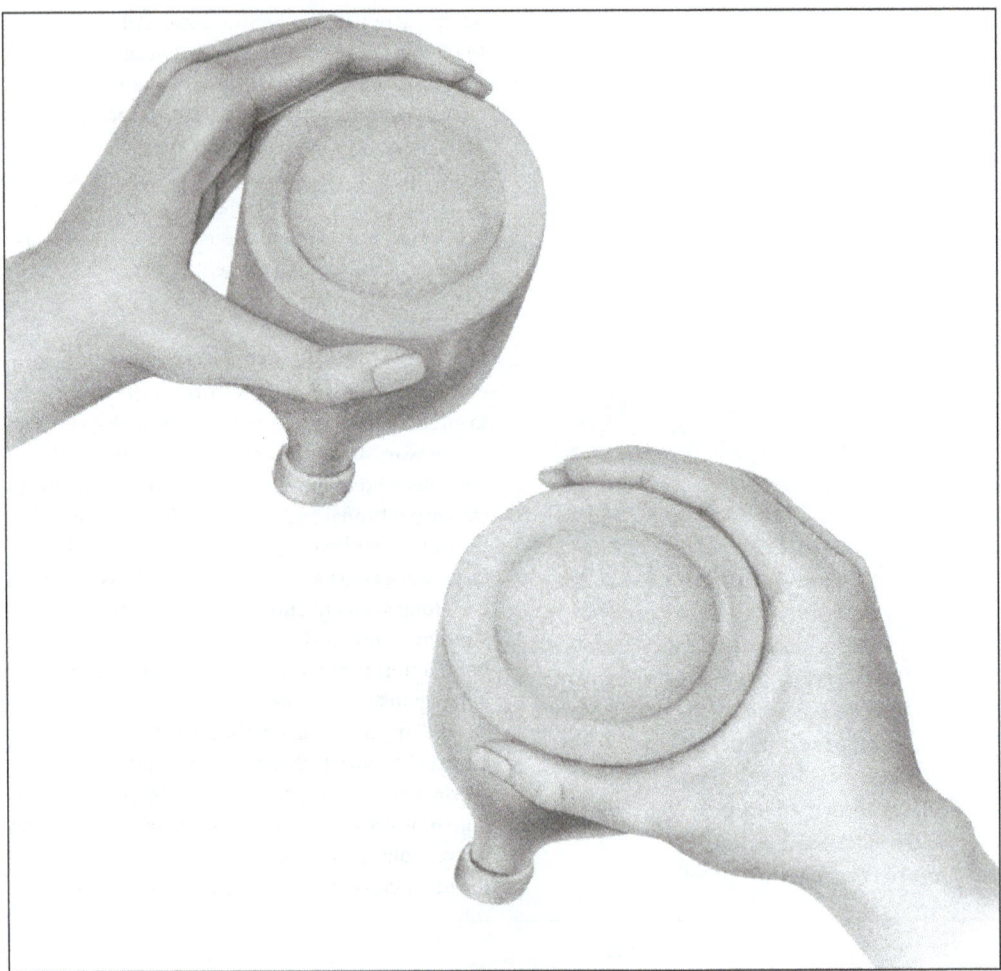

Figure 58 Positive bottle test

motor paresis in the ball of the thumb. The sensory paresis is disabling to a much lesser extent than the motor deficits, as the sensory field of the median nerve includes the most important tactile organs of the human hand. Therefore, sensory deficits must be looked for at an early stage of the illness.

Causes of carpal tunnel syndrome:

- Idiopathic from increase of the tunnel contents and narrowing of the tunnel volume during the aging process and from occupation-related strain
- Inflammatory and unspecific proliferative processes such as tendovaginitis and rheumatic diseases, metabolic diseases such as amyloidosis, goiter, myxedema, hyperthyreosis, acromegaly, and degeneration in the shunt arm in dialysis patients
- In every 10th to 20th pregnancy, usually in the last three months, increased fluid retention and increase in volume of the ulnar carpal ligament
- Space-occupying processes in the tunnel such as ganglia, fractures, strains, lipomas, bleeding
- Accessory muscles, muscle insertions, tendons, and vessels in the tunnel.

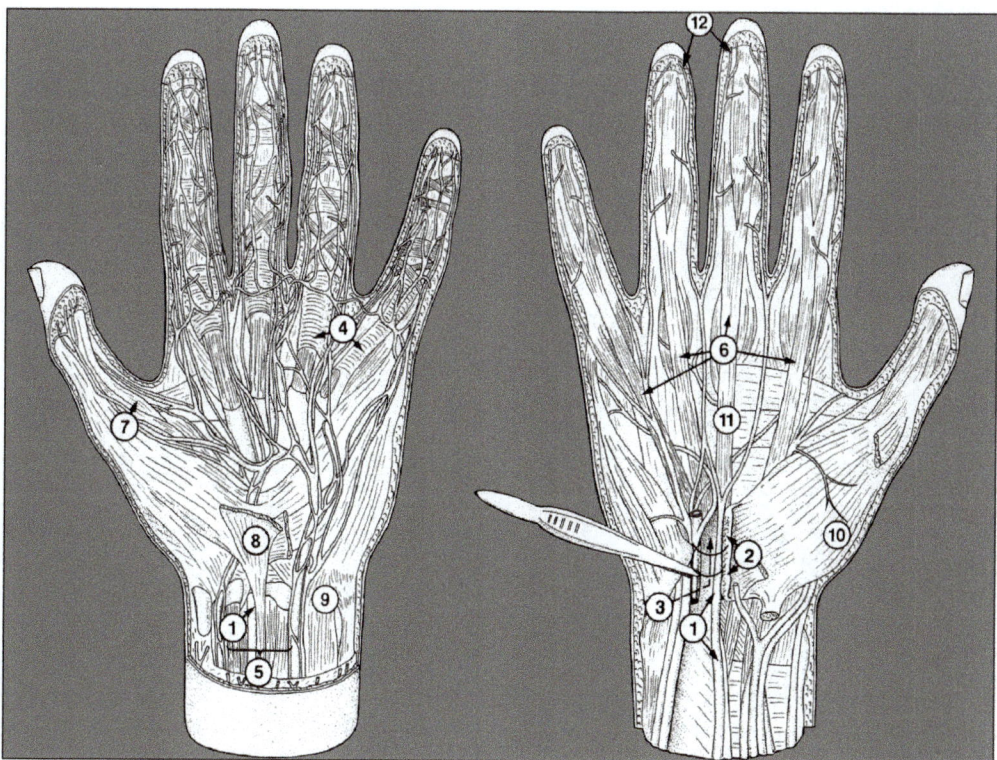

Figure 59 Schematic drawing of the hand with anatomical features of the median and ulnar nerve. 1 Median nerve that disappears under the transverse ligament in the carpal tunnel, 2 Transverse carpal ligament, 3 The two entrapment sites, loge de Guyon and carpal tunnel opened, 4 Tendon sheaths, 5 Tendons of the flexors digitorum superficialis muscles, 6 Tendons of the flexors digitorum profundi muscles, 7 Tendons of the flexor pollicis longus muscle, 8 Tendons of the palmaris longus muscle, 9 Pisiform bone without volar carpal ligament, open loge de Guyon with ulnar artery and nerve. The loge is clearly separate from the carpal tunnel, 10 Thenar branch of the median nerve to the ball of the thumb (abductor pollicis brevis muscle, opponens muscle, and flexor pollicis brevis muscle), 11 Anastomoses between the ulnar nerve and median nerve, 12 Sensory nerve fibers of the ulnar and median nerves

Diagnostics

The patient history can usually characterize the syndrome so well that it is no problem to make a diagnosis. A neurophysiological examination is always advisable. It can confirm the diagnosis, rule out important differential diagnoses, and document the extent of changes in conduction velocity, in particular to facilitate post-operative follow-up. The most important electrophysiological criterion is the reduced sensory nerve conduction velocity (NCV). Since individual fascicles can be affected in isolation, determining the NCV of a single finger nerve may not always be sufficient; several fingers innervated by the median nerve must be examined.

If the NCV also includes the little finger innervated by the ulnar nerve, a normal finding rules out polyneuropathy. The time between the electrical stimulation at the wrist up to the start of the motor response potential via the thenar enables an assessment of the motor fiber segments as distal motor latency.

Variations of the branching on the median nerve in the canal can lead to the result that only the thenar branch in the ligament is constricted. This causes pain, but the sensory NCV is unremarkable. The longer distal motor latency and a pathological EMG can verify this damage, which generally occurs with thenar atrophy. Therefore, in surgical median nerve

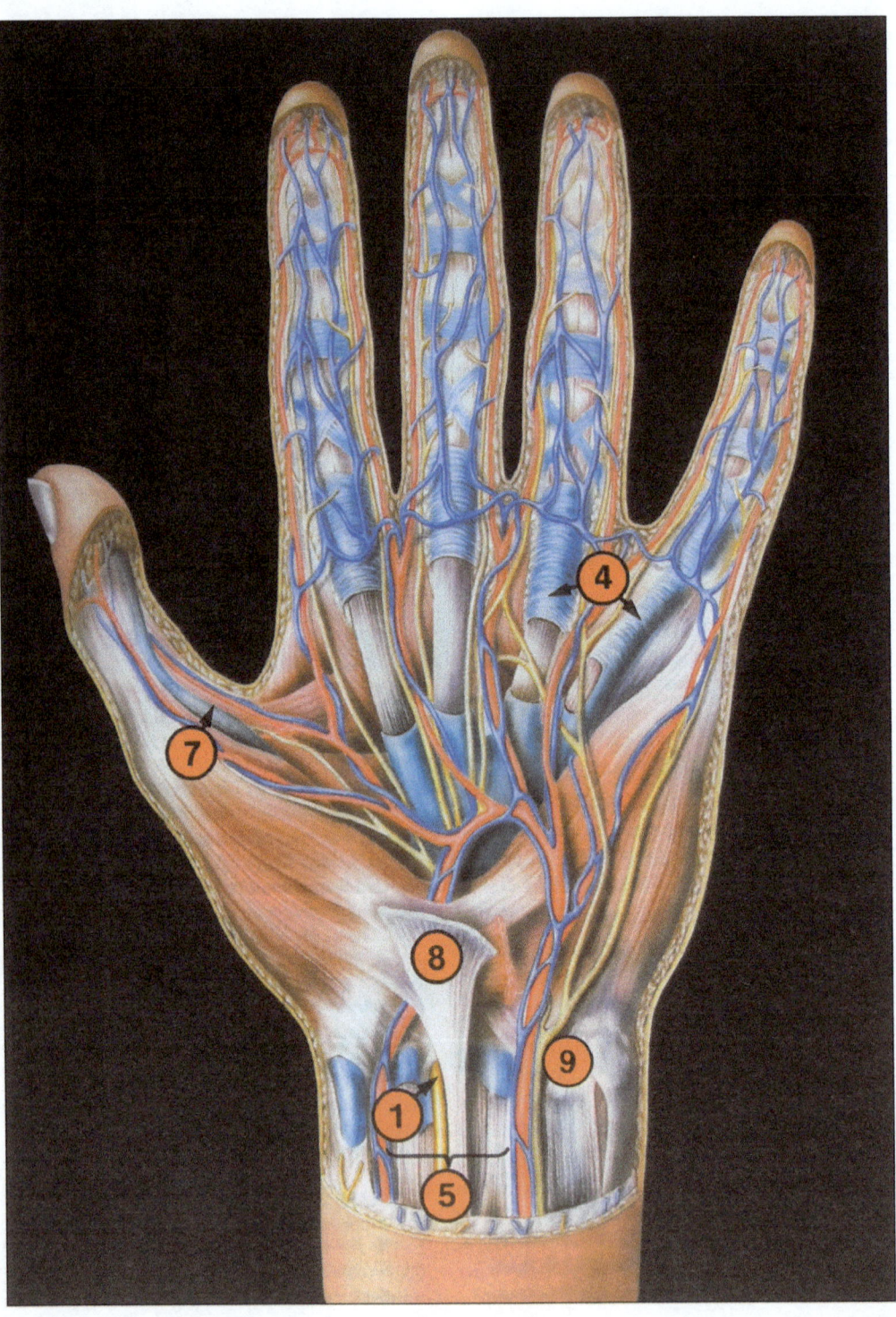

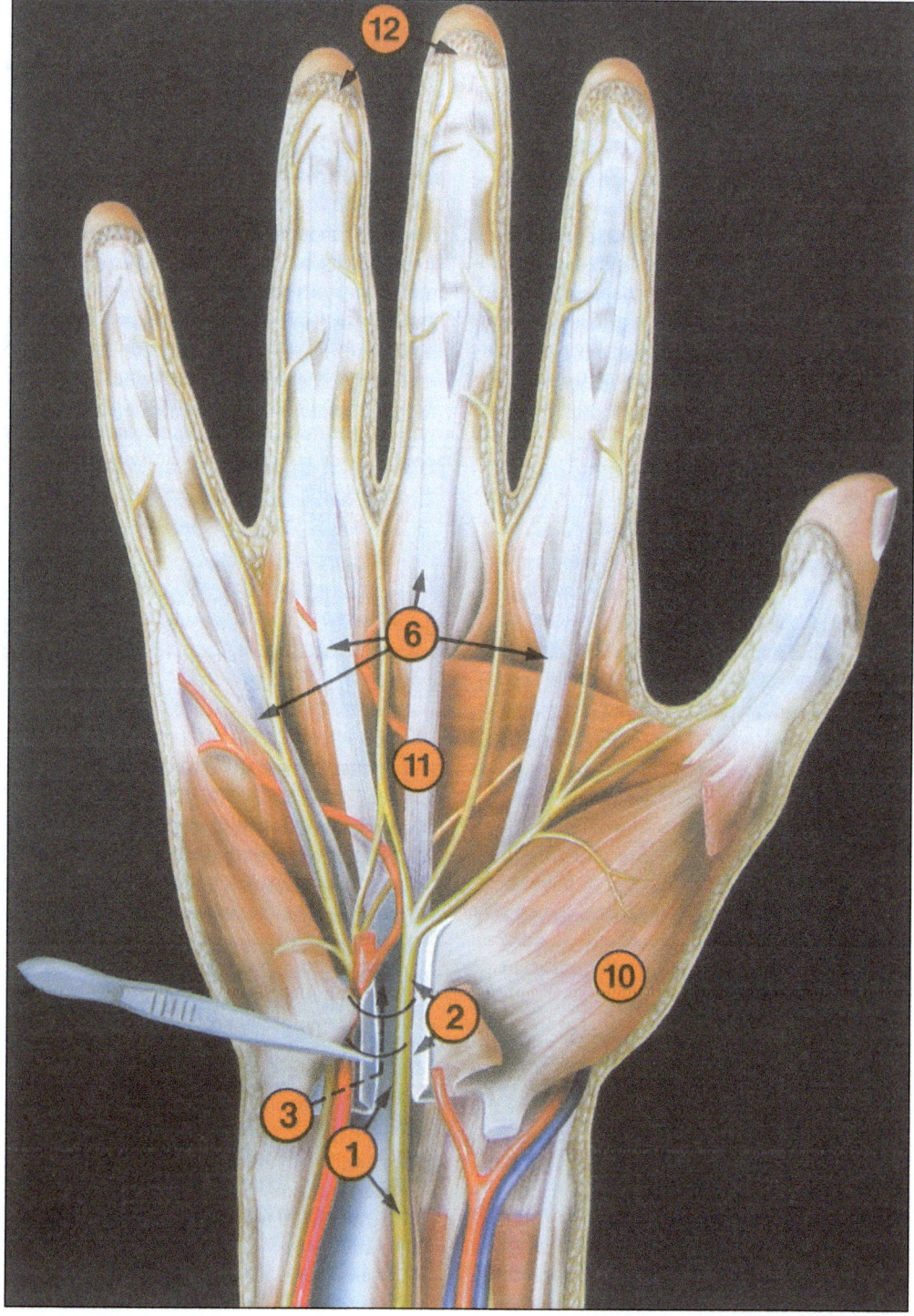

neurolysis, special attention must be given to the thenar branch.

In rare cases, motor innervation of the hand can be completely supplied by the ulnar nerve. Such cases of an "all-ulnar-hand" and other innervation anomalies must be considered in cases where the diagnosis is unclear and more specific literature should be referred to.

The differential diagnoses of a C6 syndrome and polyneuropathy should be ruled out. Pathophysiologically it is conceivable, for example, that in a C6 root lesion the trophic supply of the nerve via the axoplasmic flow is reduced, making the nerve more vulnerable, especially to compression damage. This combination of root damage and peripheral entrapment syndrome is known as a double crush lesion. A root lesion or polyneuropathy frequently triggers a carpal tunnel syndrome via this mechanism. The treatment of the underlying problem takes precedence over treatment of the carpal tunnel syndrome in such cases. Diffuse pain in the arm can also arise from plexus irritation, particularly in the thoracic outlet syndrome.

Therapy

In the stage with no objectifiable neurological deficits, immobilization of the joint during the night using a dorsal brace should first be attempted. This should prevent flexion in the wrist that compresses the nerve between the transverse carpal ligament and the flexor tendons and can trigger paresthesia from ischemia. As the next stage, corticoid injections into the tunnel can be considered. If there is no relief after two to three injections, surgical decompression should be considered. If injections are made, the risk of iatrogenic damage should not be underestimated. Blunt carpal tunnel syndromes in pregnancy are always reversible after birth and thus usually do not require any surgical treatment at this point.

Unfortunately, there is no precise guideline that gives the indication for surgical intervention for a certain clinical or electrophysiological finding. When it has been confirmed clinically that there is a carpal tunnel syndrome and deficits have been objectified by electrophysiology, the indication should be made generously.

The surgical success of neurolysis depends on the condition of the nerve, on the one hand, and on the other, whether during surgery everything was carefully explored and decompressed. However, the many variations in the course of the nerve often lead to insufficient revision.

5.10   Volkmann's contracture

Volkmann's contracture is the sequel to an ischemic circulation disorder of the forearm flexors. It occurs from lesions of the ulnar artery in supracondylar humerus fractures or repositioning attempts after fractures. As a result of ischemia, muscle necrosis and massive edema in the flexor compartments occur, resulting in the characteristic situation of a compartment syndrome. Prompt decompression by separating the fascia can prevent severe lesions to the musculature and the median nerve that proceeds through these muscles. The ulnar nerve and radial nerve are only rarely affected by such damage. The necrotic musculature is shortened and converted to connective tissue. This leads to severe contractures of the forearm, the consequences for hand mobility can be somewhat relieved only by surgical revision.

5.11   Lumbosacral plexus

The lumbar and sacral plexus are so intertwined that they form a functional unit, the lumbosacral plexus, which receives its inflow as L1–S4.

The following nerves originate in the lumbar plexus (Figure 60):
1 Iliohypogastric nerve
2 Ilioinguinal nerve
10 Lateral femoral cutaneous nerve
7 Femoral nerve
3 Genitofemoral nerve
8 Obturator nerve

The following nerves originate in the sacral plexus:
4 Superior gluteal nerve
5 Inferior gluteal nerve
6 Sciatica nerve
9 Peroneal nerve
   Posterior cutaneous nerve
   Pudendal nerve

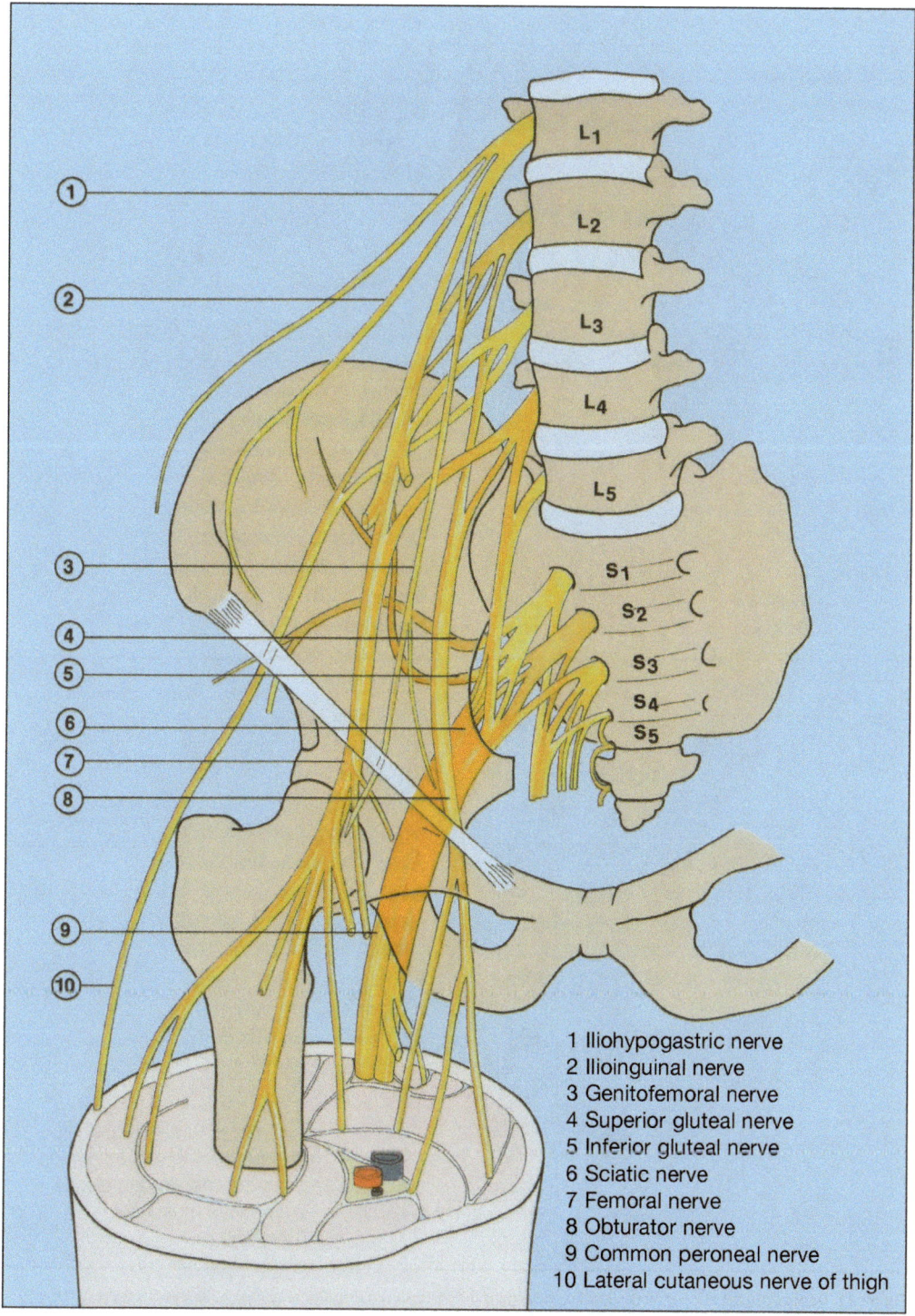

Figure 60 Anterior view of the pelvic cavity with a scheme of the anatomy of the lumbosacral plexus and nerve supply to the leg

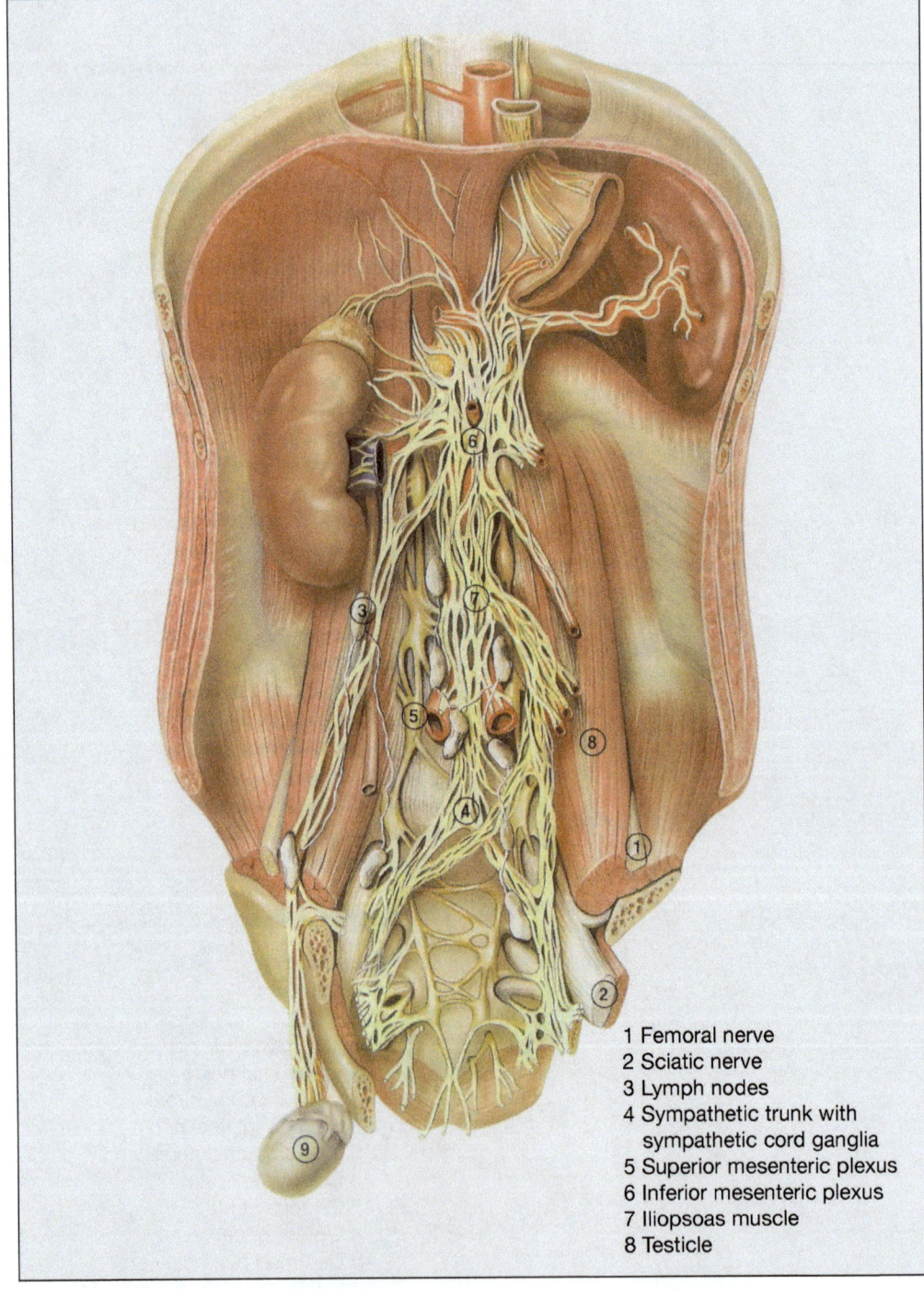

1 Femoral nerve
2 Sciatic nerve
3 Lymph nodes
4 Sympathetic trunk with
  sympathetic cord ganglia
5 Superior mesenteric plexus
6 Inferior mesenteric plexus
7 Iliopsoas muscle
8 Testicle

Figure 61 Dense sympathetic network supplies the organs in the abdominal and genital region. Pain is often the first symptom warning of a lesion

The lumbosacral plexus, surrounded by strong muscles and the osseous pelvic ring, lies protected in the retroperitoneal space. It is thus far less exposed to injury than the brachial plexus. Mechanical lesions can still affect it, however, especially with pelvic fractures. If there is compression against the pelvic bone, complicated births can lead to mechanical plexus lesions in the small pelvis. The lumbar plexus is formed from the spinal nerves between the individual strands of the psoas and iliac muscles. The major and minor psoas muscles are supplied directly from the plexus. These muscles are a typical location for spontaneous bleeding in anticoagulation treatment, so that due to their proximity to the plexus, masses caused by bleeding can lead to compression damage. It is not always easy to diagnose such damage, as severe pain, localized and radiating into the leg, often leads to a differential diagnosis of radicular damage and only secondarily to a potential plexus lesion.

In order to make a targeted diagnosis in such situations, the exact history of medication may provide the most important information. If, for example, a CT scan is ordered only on the suspicion of a herniated disc, a small CT diagnosis window must be set only including the spinal canal in order to improve resolution. But in that small window, retroperitoneal bleeding can easily be overlooked. Spiral CT gives quick orientation as to the situation; a muscle hematoma can be seen as an impressively large mass. The fact that surgery will hardly bring relief for such a plexus lesion can also be seen from the CT finding, as the hematomas are not seen in isolation, encapsulated in the muscle, but as a blood-saturated swelling of the entire musculature. In this situation, the most urgent therapeutic goal is to prevent further hemorrhaging. Further conservative therapy is also supported by the experience that the hematomas are spontaneously resorbed and the diminishing mass usually leads to the complete restitution of nerve function.

With space-occupying masses, an abscess must always be considered as a possible differential diagnosis. In the iliopsoas muscle, such processes can continue downward and appear as swelling in the groin (Figure 65).

Due to the proximity of the plexus and in particular the sympathetic trunk to the lymphatic vessels, damage frequently occurs in association with lymphoreticular or metastasizing illnesses (Figure 61). Generally, the patient complains of rather unspecific symptoms, such as more or less long-term diffuse, sciatica-like or lumbago-like pain. Sometimes the deficits have a radicular character. At this stage, observation and testing sweat secretion can sometimes verify that peripheral nerves are affected and point toward the correct diagnosis.

The iliohypogastric and ilioinguinal nerves correspond to the intercostal nerves, and they are also involved in the motor innervation of the abdominal wall musculature. The iliohypogastric nerve provides sensory innervation from the suprapubic area to the hip (Figures 13a, 13b). The ilioinguinal nerve continues in the inguinal canal and can be damaged in herniorrhaphy procedures. Pain then radiates into its sensory supply territory, which includes the pubic and genital area. The genital region is also supplied with sensory innervation by the genitofemoral nerve.

5.11.1 Diabetic plexopathy

Patients with longstanding diabetes can develop a plexopathy during therapy adjustment or after losing a great deal of weight, with a peak between ages 60 and 70. The syndrome is also called diabetic amyotrophy or proximal diabetic neuropathy. The exact cause is unknown, but diabetic microangiopathy of the vasa nervorum is assumed. Suddenly, but also gradually, and often at night, there is severe pain in the back, hips, or thighs. This is accompanied by paresis of the pelvic girdle musculature. However, the sensory deficits are usually slight. If the quadriceps muscle is very weak, there may be suspicion of an isolated femoral nerve paresis. However, more precise clinical and electrophysiological examinations show that in addition to the mononeuropathic involvement, there are also other disorders typical for a plexopathy. Occasionally, these symptoms also occur idiopathically without any predisposing illness.

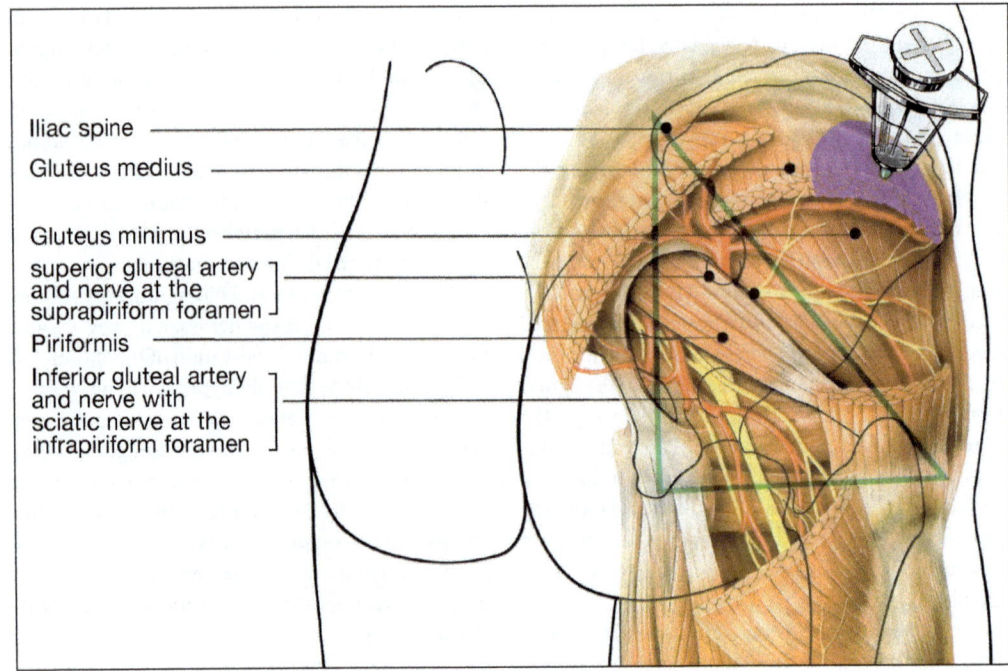

Iliac spine

Gluteus medius

Gluteus minimus

superior gluteal artery and nerve at the suprapiriform foramen

Piriformis

Inferior gluteal artery and nerve with sciatic nerve at the infrapiriform foramen

Figure 62 Orientation aid for intragluteal injection. Green = Critical area. Violet = The needle is inserted into the gluteus medius and minimus vertical to the surface of the body

5.11.2 Gluteal nerves, injection paralysis

The segments of roots L4–S3 form the sacral plexus, from where the sciatic nerve originates. This nerve proceeds at the level of the piriformis muscle out of the small pelvis through the infrapiriform foramen. It is accompanied by the inferior gluteal nerve, which innervates the gluteus maximus muscle. The superior gluteal nerves proceed through the suprapiriform foramen and innervate the gluteus medius and minimus muscles. These muscles are responsible for the horizontal stabilization of the pelvis. Their paresis leads to a gait with swaying, swinging pelvis and trunk movements, as often seen in patients with muscular dystrophy in the pelvic girdle. Due to paresis of the gluteus medius muscle in the standing leg, the muscle cannot stabilize the pelvis while standing on one leg, and the pelvis tilts and sinks on the contralateral side. This test and its result are known as positive Trendelenburg's sign.

Injection paralysis occurs from:
1. Directly injecting a vessel, a nerve, or a nerve root
2. The neurotoxic effect of injected substances
3. The space-occupying effect of a depot

Not infrequently, faulty intragluteal injection can damage the sciatic nerve or the superior gluteal nerve. This can be prevented by the correct injection technique into the outer quadrants of the gluteal musculature (Figure 62). Aspiration before the injection ensures that the tip of the needle is not in a vessel, in particular not in an artery. Injection into an artery can cause extensive muscle necrosis. Intramuscular injections should be avoided for anticoagulant therapy in any case, as the injection can lead to extensive hematomas with nerve compression.

In injection paralysis, the symptoms are immediate pain if the injection hits a nerve, or with a short delay if the medication damages the nerve. Injection lesions can also occur with paravertebral injections

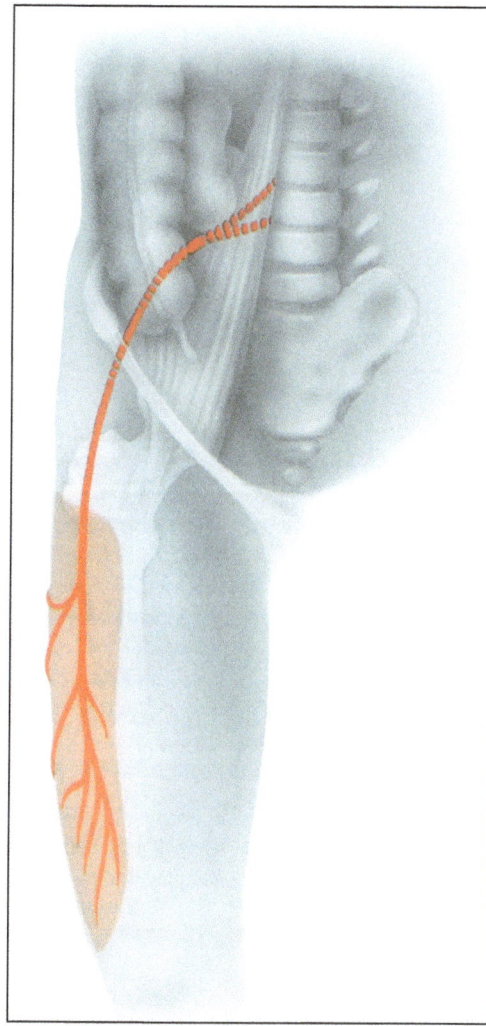

Figure 63 Course and sensory skin area supplied by the lateral cutaneous femoral nerve

the injection. In such cases, it is advisable to document the injection site, the injected substance, and patient's reaction exactly and inform the liability insurance.

5.12 Cutaneous femoral nerve

Of the spontaneously occurring peripheral nerve lesions in the leg, that of the cutaneous femoral nerve is the most frequent; it is also known as meralgia paresthetica. The nerve must move through the inguinal ligament as it exits the pelvis; an entrapment syndrome can occur there. Behind the inguinal ligament, the nerve proceeds downward at almost a right angle, and here, in particular fat, hanging abdominal walls exert traction. The exit site of the nerve near the anterior inferior iliac spine is tender in meralgia; a local blockade can often be useful for diagnosis and therapy. Clinical symptoms of meralgia are burning pain and sensory disorders on the outer side of the thigh (Figure 63) that can be provoked by extending the hip joint with the reverse Lasègue maneuver (Figure 33). Walking and lying are frequently painful, patients often use sitting as a relief position. The radiation pattern of pain is similar to an L3 syndrome. But a lack of paresis or reflex deficits in the quadriceps does not support this.

5.13 Femoral nerve

The nerve originates in the segments L2–L4, is formed between the iliopsoas and iliac muscles, proceeds lateral to the femoral artery under the inguinal ligament to the quadriceps muscle and innervates it (Figure 64).

Below the inguinal ligament, it gives off the cutaneous branch that innervates the ventromedial thigh. As a purely sensory saphenous nerve, it continues in distal direction and innervates the medial knee and the inner side of the lower leg to the ankle.

Clinical symptoms for a high femoral lesion are, in addition to numbness at the inner side of the leg, severe limping due to the loss of the quadriceps as knee extensor. Climbing stairs is impossible; the patella reflex is absent.

where the local anesthetic contained in the mixture is often responsible for additional complications. Injecting into the cerebrospinal fluid space directly or via the root sheath frequently causes – usually transient – paraplegic lesions. A major cause for the damage of an injection lesion is the toxic injection substance, which is tolerated only by muscle tissue and may therefore not be injected into any other tissue. The unphysiological pH level can quickly denature the protein in the nerve, so that this process alone makes nerve damage irreversible soon after

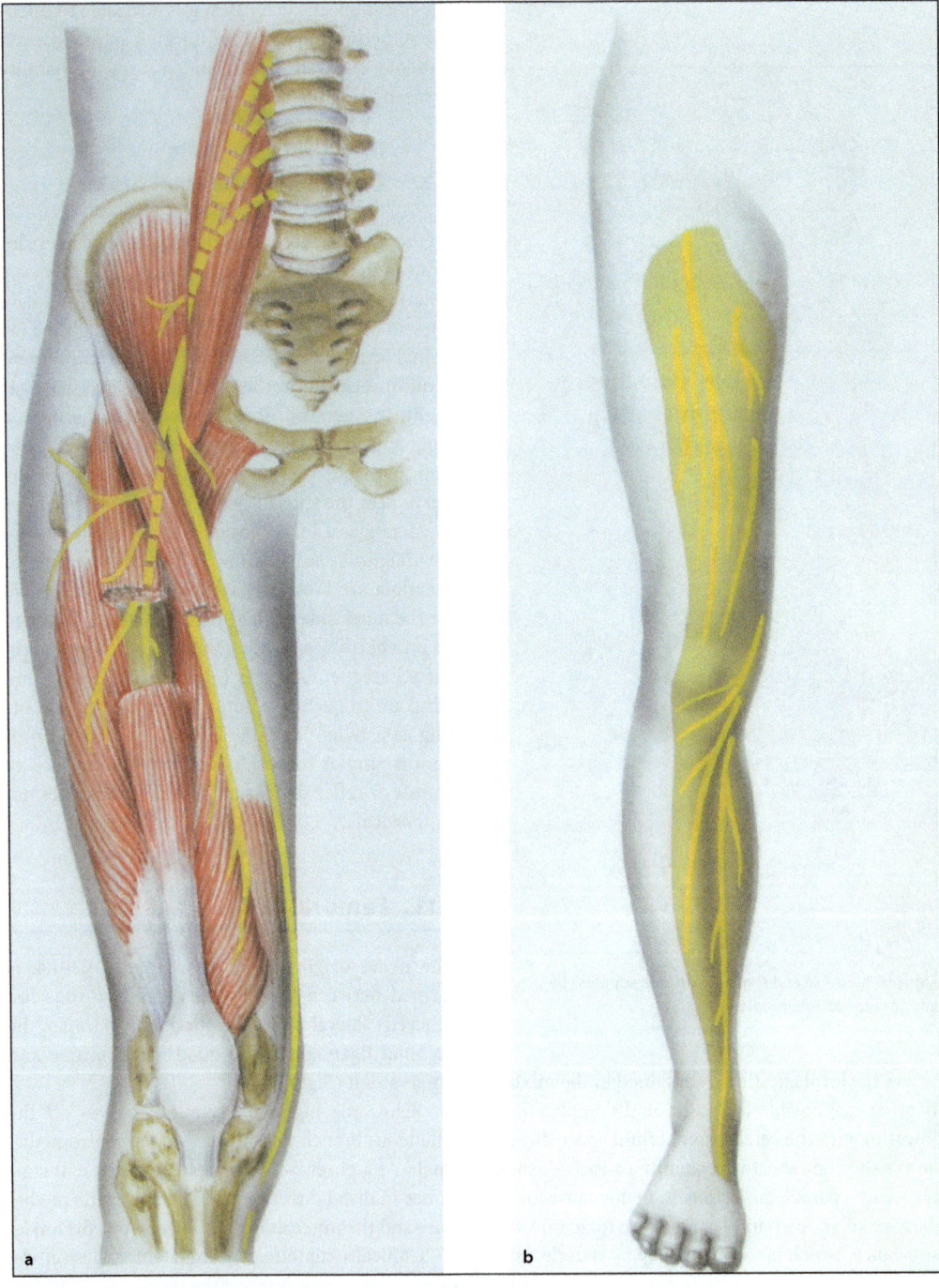

Figure 64 Course of the femoral nerve with (**a**) muscle and (**b**) skin supply

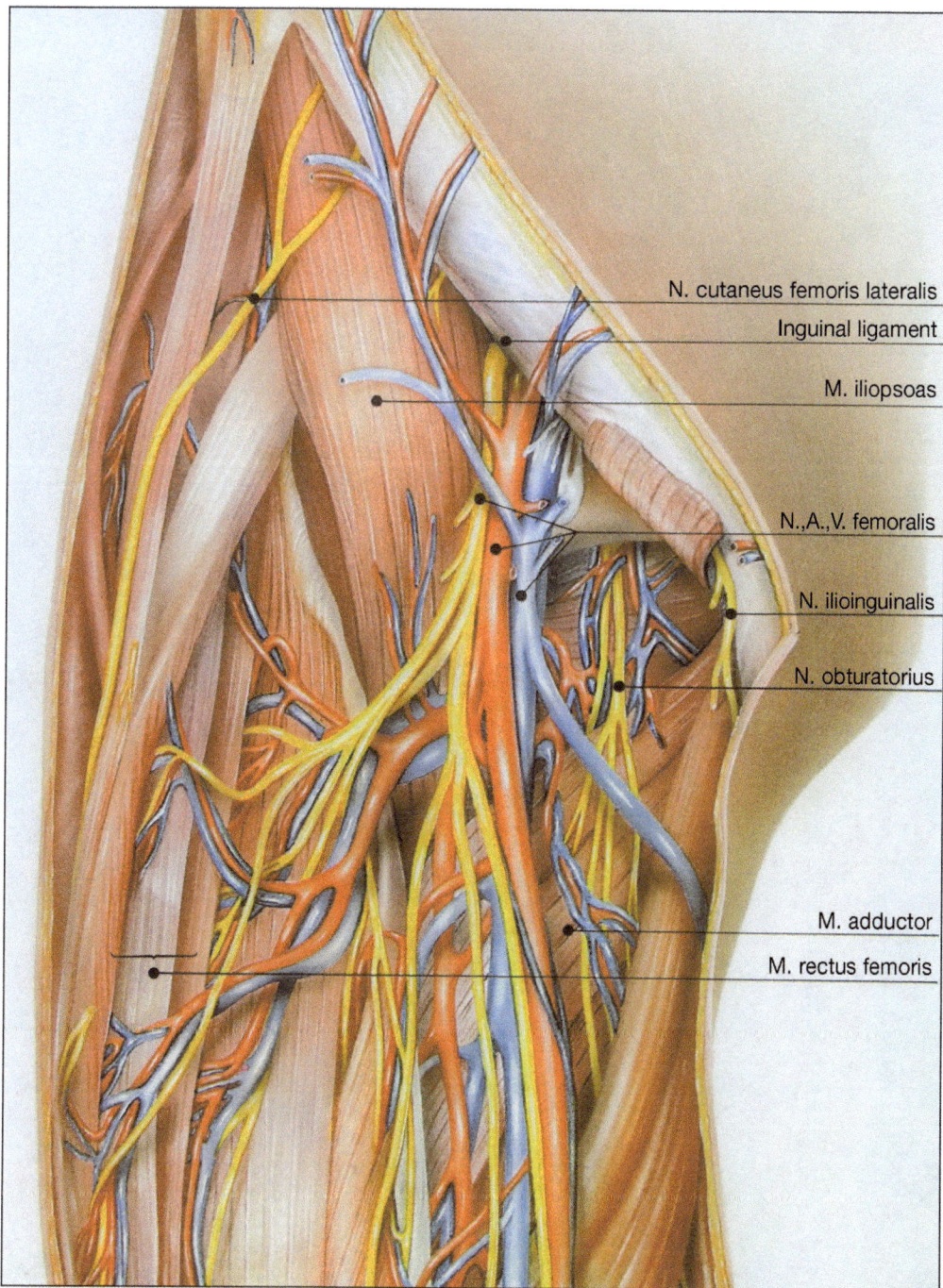

Figure 65 Anatomy of the inguinal region

N. cutaneus femoris lateralis

Inguinal ligament

M. iliopsoas

N.,A.,V. femoralis

N. ilioinguinalis

N. obturatorius

M. adductor

M. rectus femoris

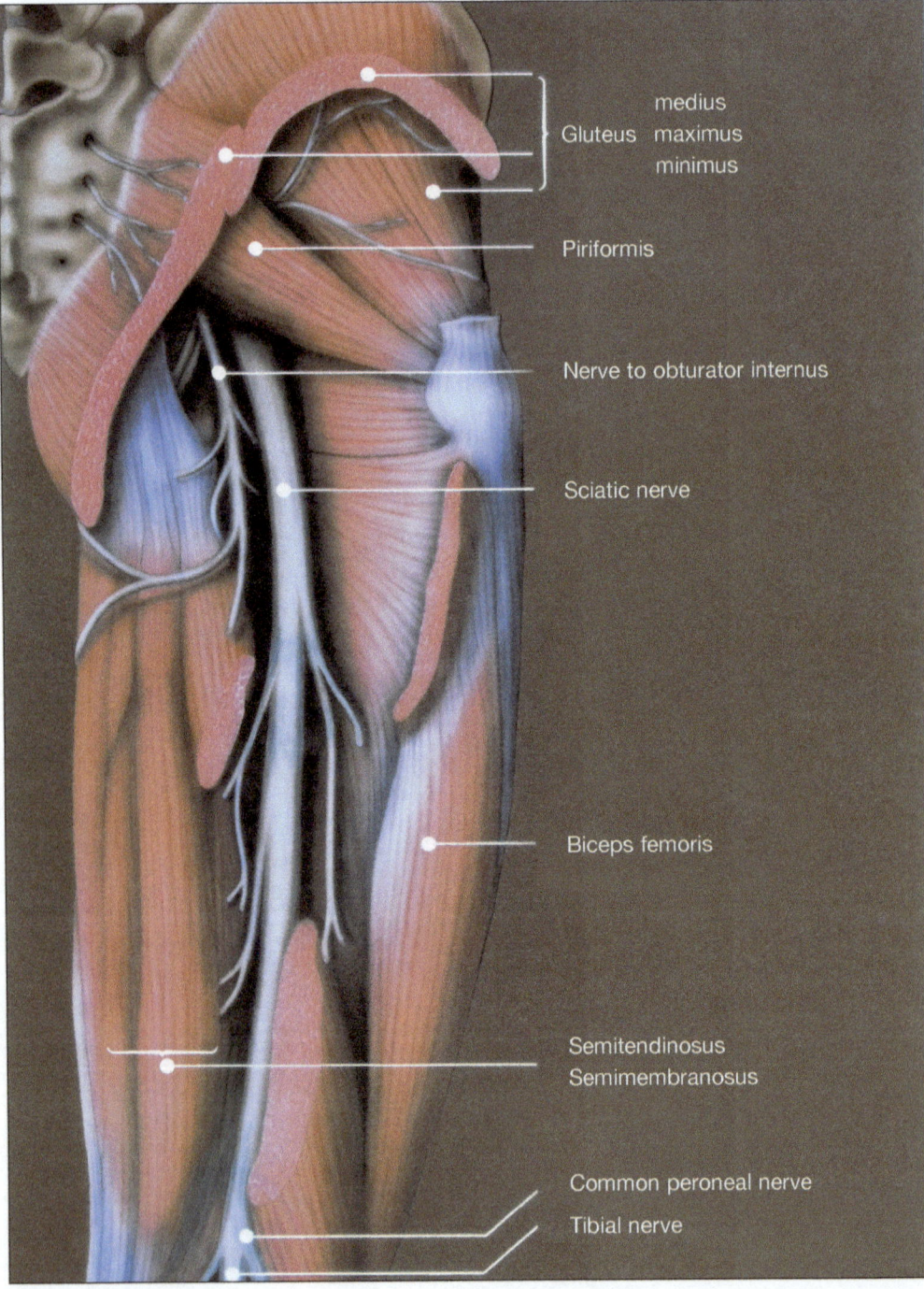

Figure 66 Course of the sciatic nerve in the thigh

Damage to the femoral nerve is usually the result of direct trauma such as stab wounds or arterial punctures or compression in the groin. Intra-abdominal procedures rarely cause damage, but it can occur during implantation of artery grafts, hysterectomy, or lymph node extirpation. Damage due to positioning occurs in the lithotomy position by bending the nerve and the compression this causes at the inguinal ligament (Figure 65). Conversely, traction trauma can be caused by hyperextension at the hip in sports or from faulty positioning in coma. Surgical manipulations in bone operations of the femur or fractures can cause extension trauma; in a hip replacement, damage to the femoral nerve is also not infrequent. Isolated damage to the femoral nerve can occur with diabetic mononeuropathy multiplex. More exact examinations and damage verified in areas not innervated by the femoral nerve then often draw attention to a plexopathy.

As a differential diagnosis, an L3 lesion must always be considered.

Saphenous nerve lesions are infrequent despite the long course of the nerve near the surface. Damage is usually caused in surgical procedures such as vein stripping or vein removal.

5.14 Sciatic nerve

Sciatica pain is an irritation that primarily affects the radicular inflow to the sciatic nerve. Pain and paresis are projected to the territory supplied by the sciatic nerve corresponding to the irritated segment fibers (for more information see chapter 4.7 "Lumbar root syndromes" and chapter 9.5.4 "Low back pain, sciatica").

The sciatic nerve originates from the lumbosacral plexus with L4–S3 fibers forming the largest part. It continues through the infrapiriform foramen and proceeds well protected under the femoral biceps to the knee (Figure 66). There are many anatomical variations of the course. Either the individual fibers combine to one nerve at the level of the piriformis muscle and do not separate until the back of the knee or the tibial and peroneal segments of the nerve are already separated below the pelvis. The peroneal segment can even proceed through the piriformis muscle and suffer an entrapment syndrome there. The semi-

membranosus, semitendinosus, adductor magnus, and long head of the biceps muscles are supplied from the tibial segment. The peroneal segment supplies only the short head of the femoral biceps muscle.

A sciatic nerve lesion affects nearly all thigh flexors as well as all lower leg and foot muscles. The damage is usually caused by trauma or compression lesions from masses in the small pelvis. But some habits such as sitting too long on the toilet reading the newspaper can also lead to paresthesia. If this continues long enough, compression lesions can develop, in which the relatively small contact area of the toilet seat damages a circumscribed area in the thigh nerves. Sciatic paresis such as this can also occasionally be caused by intoxication, when the patient lies in a position damaging the nerve for too long. The patient in this condition does not feel the warning symptom of the paresthesia and pain triggered by the compression damage that would normally lead him to change position and relieve the pressure on the nerve.

Traction trauma, especially in accidents, often affects the nerve in the thigh; but clinical deficits are restricted to the innervation territory of the peroneal or tibial nerve. The peroneal segment of the sciatic nerve is more prone to lesions than the tibial segment.

5.15 Tibial nerve

The nerve contains fibers from segments L4–S3, proceeds in the sciatic nerve to the back of the knee and through the tendinous arch of the soleus muscle, well protected under the calf muscles, to the foot (Figure 67). Lesions of the nerve in the sciatic segment are identified by paresis in the thigh flexors – the semimembranosus, semitendinosus, and long head of the femoral biceps muscles.

The tibial nerve provides motor innervation to the calf and foot flexor musculature. The nerve is rarely damaged in the lower leg area except in fractures and knee disorders.

Sensory innervation of the lower areas of skin over the calf is provided by the tibial nerve. It anastomizes with the lateral cutaneous nerve from the peroneal nerve to form the sural nerve, which provides sensory innervation to the skin over the outer

5

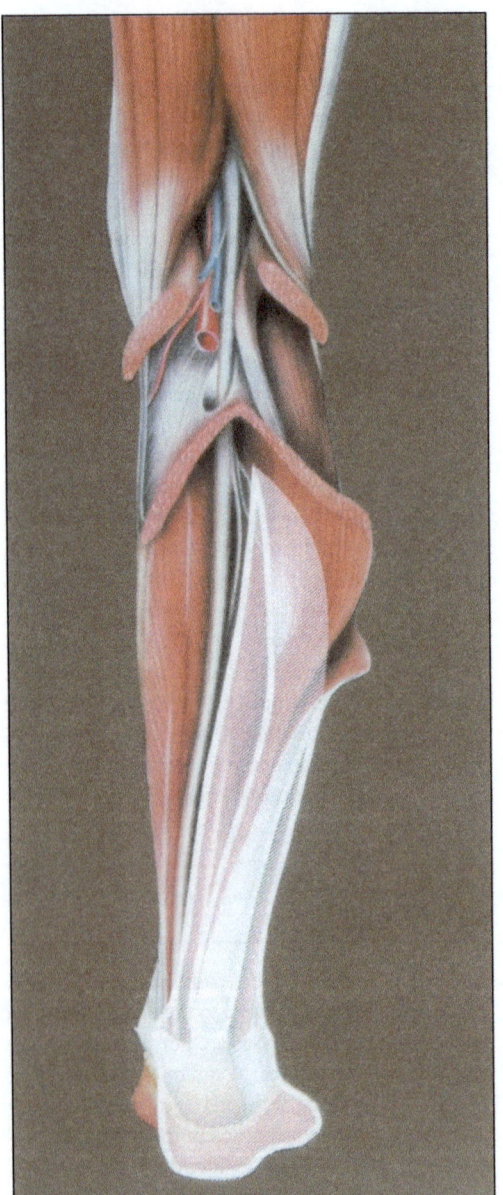

Figure 67 Course of the tibial nerve in the thigh, the white hatched area shows its sensory supply area

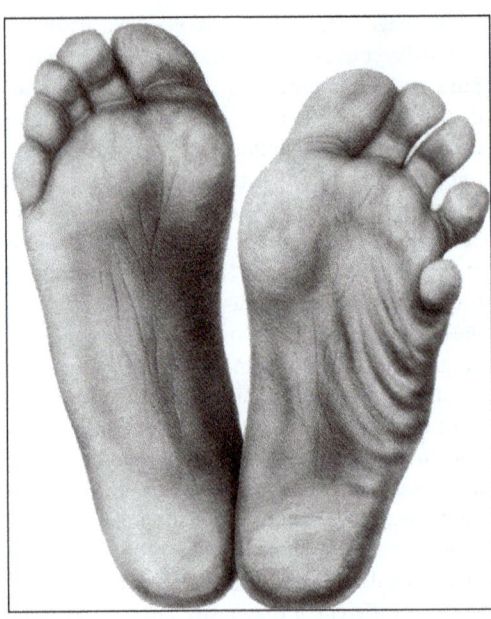

Figure 68 Tarsal tunnel syndrome with paresis of the flexor muscles of the right foot

Tarsal tunnel

Muscle tendons proceed from the calf in a curve around the lateral malleolus of the ankle and insert at the medial edge of the foot. For this, the top of the curve proceeds in a ligament to the fixation site. The distal branches of the tibial nerve, the medial and lateral plantar nerves, proceed through this connective tissue pathway. This region of the tarsal tunnel can entrap the nerves. Clinical symptoms are unpleasant paresthesia in the ankle and sole of the foot and paresis of the small foot flexors (Figure 68).

Fractures, dislocations, or tight shoes can cause changes to the connective tissue in the tunnel. If there is unexplained pain in the foot, tarsal tunnel syndrome is often diagnosed for lack of a better idea, and unnecessary surgery is performed. If the symptoms include only pain, the pitfall-ridden electrophysiological diagnostics are not very useful in any case. Diagnostic blockades with a local anesthetic can give more information. Verifying a sweat secretion disorder can objectify a lesion.

Morton's metatarsalgia

Small neuromas of the plantar nerves are assumed to be the cause of localized burning pain occasion-

ankle joint and the outer edge of the foot. Sural nerve lesions are usually iatrogenic, caused by biopsies. Since extirpating the nerve at the lower part of the lower leg produces very few sensory deficits, it is suitable for removal to diagnose polyneuropathies.

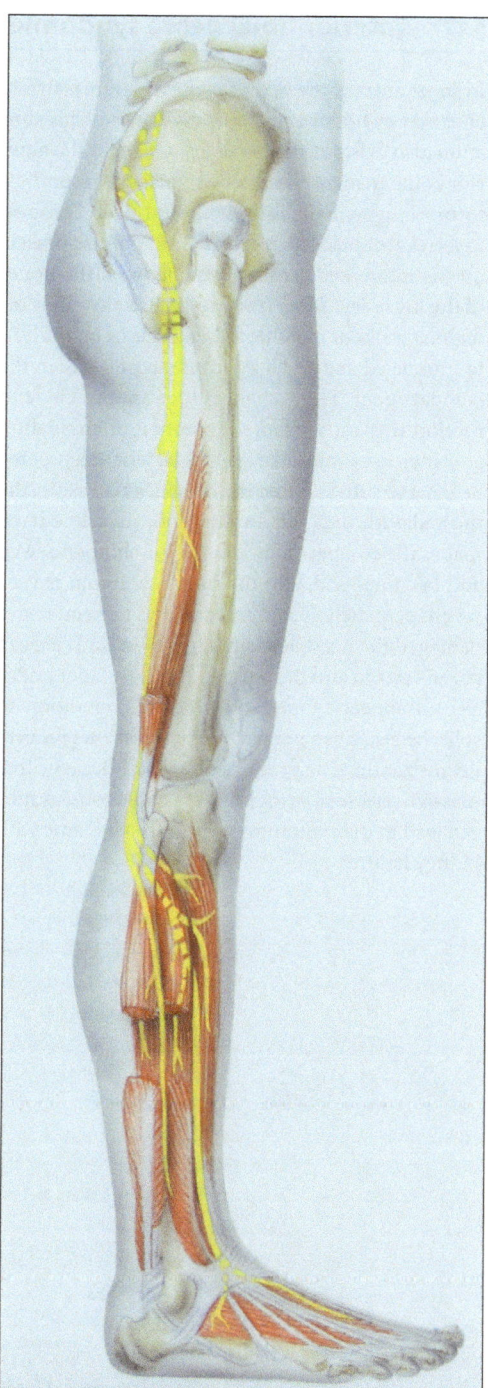

Figure 69 Course of the peroneal nerve with its motor and sensory innervation areas

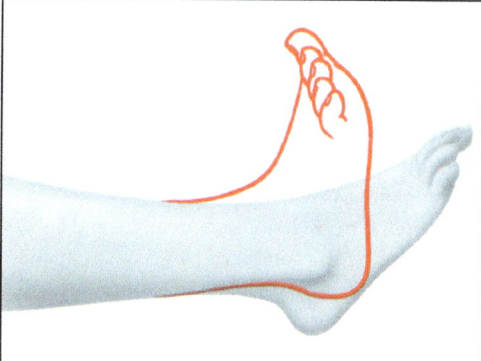

Figure 70 Function test of the deep peroneal nerve. The primary dorsiflexor muscle is the anterior tibialis, which cannot normally be overcome in the strength test. The extensor hallucis muscle lifts the large toe, the extensor digitorum protrudes clearly at the lateral dorsum of the foot when contracting and lifting the toes

ally felt in the sole of the foot. The symptoms are initially related to strain, later becoming permanent. Shoe inserts that relieve the heads of the metatarsal bones, the site where the nerves are usually damaged and neuromas develop, can be helpful. If this therapy is not effective, steroid injections can be attempted before surgery is undertaken.

5.16   Peroneal nerve

The nerve originates at the sciatic nerve and proceeds from the back of the knee around the fibula head to the peroneal compartment. Before this, it gives off a skin branch that supplies the outer side of the lower leg. In the peroneal compartment, it divides into two branches. The superficial peroneal nerve innervates the following muscles at the outer side of the lower leg, which act as pronators in the foot (Figure 69):

- Peroneus longus muscle
- Peroneus brevis muscle

The deep peroneal nerve supplies the following foot and toe dorsiflexor muscles in the lower leg:

- Tibialis anterior muscle
- Extensor hallucis longus muscle
- Extensor hallucis brevis muscle

- Extensor digitorum brevis muscle at the top of the foot

The deep peroneal nerve continues as a sensory branch to the dorsum of the foot, where it innervates the space between the first and second toe as an autonomous area.

Peroneal nerve lesions are frequent and are easily confused with a combined L4/L5 symptomatic or central foot dorsiflexor paresis. Common to all is the typical peroneal gait. The tip of the foot cannot be lifted from the ground in the swing phase of the leg and drags on the ground (Figure 70). There are no reflex disorders in peripheral peroneal nerve lesions, and the posterior peroneal muscle innervated by the L5 and tibial nerve is always intact in foot supination. However, gradually progressive dorsiflexor paresis with no sensory disruption can also be a symptom of dystrophic muscular diseases.

The level of the lesion of the peroneal nerve is often impossible to determine clinically, as peroneal nerve lesions can also be caused by damage located high on the sciatic nerve. In such cases, an EMG examination of the short head of the femoral biceps can be helpful. But as in high lesions, not all fibers of the peroneal nerve are necessarily damaged, patterns of lesions located at lower locations can be imitated.

The peroneal nerve is most frequently damaged in the area around the knee. That is where it emerges from the back of the knee and proceeds close to the surface, can be easily felt under the skin, and is thus easily damaged around the fibula head. However, before it enters the musculature that it supplies, it divides into a deep and a superficial branch. The clinical symptoms depend on whether the nerve was damaged at the main stem or only at one of its branches, with corresponding function deficits of musculature and sensitivity.

Damage in the back of the knee and at the fibula head are usually caused by poor positioning, for example, from habitually crossing the legs while sitting or sleeping, from faulty positioning of unconscious persons, too tight or poorly fitting casts, from garters, or from working in a kneeling or squatting position.

5.17   Anterior tibial nerve syndrome

In some areas of the body, the musculature is firmly enclosed by bones and connective tissue. This situation also exists at the lower leg, where in the anterior compartment the anterior tibial, extensor digitorum longus, hallucis longus, and deep peroneal muscles are enclosed in tough, fibrous muscle fascia on the interosseous membrane between the bones of the lower leg. Bone fractures, muscle swelling after long walks or marches with muscle ischemia lead to muscle edema and a rapid rise in pressure in the compartment. These changes set a vicious circle in motion that causes more disruption of circulation and increases muscular ischemia. The deep peroneal nerve suffers under the increasing pressure, the musculature dies off. In addition, muscle edema makes the compartment next to the shinbone swell and become hard, and the initial symptom is very severe pain. When compression and ischemia have damaged the nerve and muscle, motor and sensory paresis sets in and the pulse in the dorsal foot artery also disappears. Only an immediate operation to split the fascia can prevent the progression of severe necrotization. The products of muscle decay with a massive release of myoglobin are dangerous, as they can lead to the symptoms of a crush syndrome with kidney failure.

Acute and chronic pain

Acute or chronic pain is a complex multidimensional neuronal response that draws attention to damaging or potentially damaging situations. A complex cascade of events takes place between the triggering occurrence of irritation and the final result of pain perception. In the beginning the nociceptor transforms the noxious threat into a signal. Then the nociceptive signal is transduced to the central nervous system, where synaptic plasticity takes place. The noxious signal undergoes modification by regional facilitation or inhibition influenced by ascending and descending impulse patterns. The last step is pain perception. It involves sensory-discriminative, affective, and emotional experiences of the patient's present and past. The experience is subjective and modulated by genetic disposition as well as by factors such as cognition (e.g., distraction, fear, anxiety, and catastrophizing), mood, and beliefs.

Pain can be divided into acute and chronic pain. Pain is regarded as chronic when it lasts for more than 3–6 months. This definition is used in many diseases when chronicity is being defined. It is a paradigm strictly relying on the time but without any concern for the pathomechanisms. In this context a better definition for chronic pain is: a persisting pain after healing time and lacking the acute warning function of physiological nociception. In 2015 the International Association for the Study of Pain (IASP) suggested to introduce a classification for chronic pain in the International Classification of Diseases (ICD). In 2018 the upcoming and revised ICD-11 edition will be published by the World Health Organization. The IASP proposed relevant pain disorders and classified them according to the perceived location, etiology, or primary affected anatomical system:

- Chronic primary pain
- Chronic cancer pain
- Chronic posttraumatic and postsurgical pain
- Chronic neuropathic pain
- Chronic visceral pain
- Chronic musculoskeletal pain

In acute pain, tissue irritation occurs, for example, in the skin (Figure 71), and within the somatosensory system, the damage or the potentially damaging irritation is recognized, processed, and a defense or healing reaction is initiated by the nociceptive system. When the triggering noxa are removed, the pain disappears and the event ends on its own. Pain is a fundamental and important biological warning signal and, as such, indispensable. Without the pain sensation, the integrity of the organism cannot be ensured and it is thus one of the most important life-saving functions. This significance of pain makes it easy to understand why there are so many mechanisms for conducting pain. What can be problematic for therapy is the variety of signal pathways arranged usefully by nature as a safety mechanism for ensuring that pain is still perceived even if the normal pain transmission pathways are inactivated. Pain first draws attention to protect the organism in the event of an illness-causing event and, ideally, the patient can be healed and thus relieved of the pain as well. The situation is completely different when the nociceptive system itself is affected and damaged. In this case, there is a pathobiological disorder of nociceptive impulse generation and impulse activity and their processing in the central nervous system. Now the pain has become an independent and chronic illness. It has lost its warning function and must be viewed as a pain disease of its own. Therapy for chronic pain can be effectively and usefully planned only when the conditions of the individual pain situation are analyzed and mechanisms and the circumstances of the development of pain are taken into consideration.

6.1   Pain concepts over time

Historically, pain was believed to be a simple effect-response reflex pattern. This thought stemmed from the influential pain paradigm by René Descartes (1596–1650). In the American Civil War, many patients had wounds with concomitant damage of peripheral nerves and they experienced excruciating pain. The reasons for the development of this pain after nerve injuries were the subject of various studies. For example, at that time Silas Weir Mitchell described the clinical picture of causalgia. Motivated by the agony caused by these symptoms, the English neurologist Henry Head began his own research into the phenomena of nerve regeneration after his sensory radial nerve was severed. He came to the conclusion that there must be two different

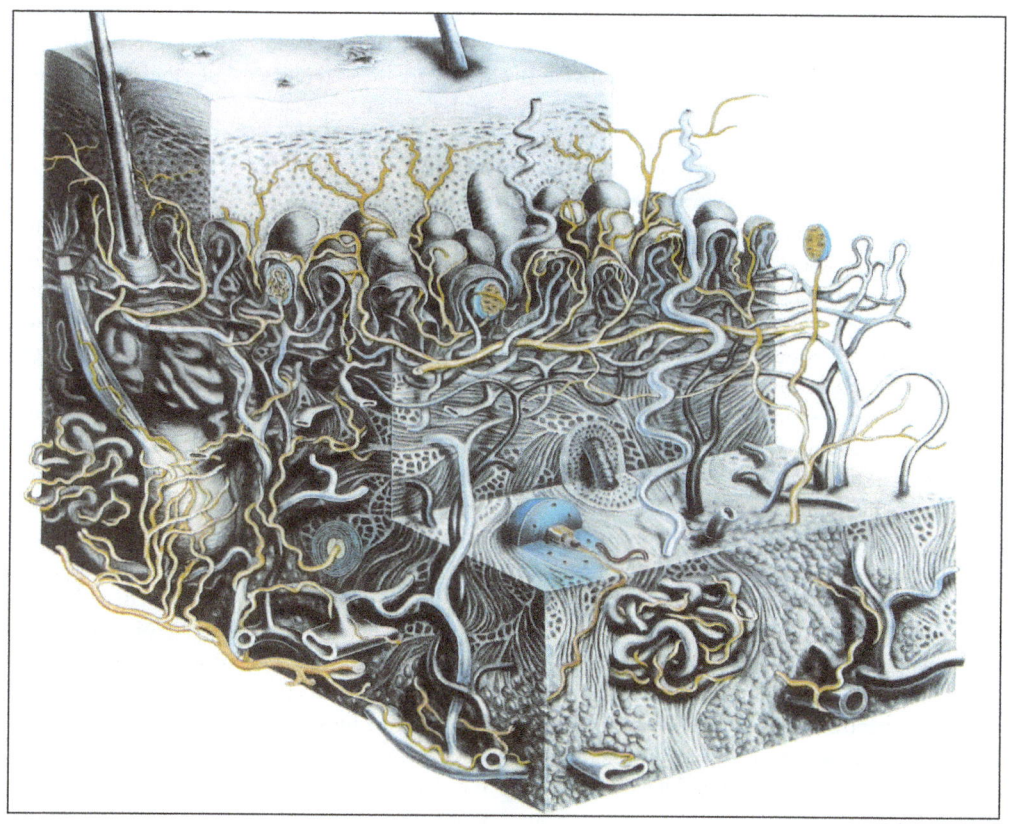

Figure 71 Innervation of the skin. Skin, muscles, bones, ligaments, joints, and organs are transversed by a dense network of nerve fibers (yellow) that register nociceptive events

systems of cutaneous receptors and conducting nerves. One kind of nerve was responsible for primitive (pain, extreme differences in temperature and yielding ungraded, diffuse all-or-nothing impressions) and thus protopathic sensation, the other kind for specific and thus epicritical sensation (touch, two-point discrimination and temperature changes). In his perception, epicritical sensitivity overlaid the protopathic type. Assuming a competitive mechanism, it was considered how the phenomenon of pain could be explained by stimulus and inhibitory phenomena. On the basis of these old ideas, in 1965 Melzack and Wall suggested a theory for the processing of nociceptive information in the spinal cord. They assumed at that time that there were two competing information systems influencing the dorsal horn. One system included afferents from thick, non-nociceptive fibers, the

other nociceptive information from thin fibers. However, there is a gate, the substantia gelatinosa, positioned between the two systems. Before this, presynaptic inhibition via interneurons had been discovered by John Eccles, presumed to be in the substantia gelatinosa. Information from the thick fibers can affect the gate so much that the influx of information from the thin fibers to the spinal cord is prevented. The gate control theory hypothesizes that a painful stimulus is first conducted directly to the dorsal horn via thin fibers. If pressure, tactile, or other non-nociceptive receptors on the skin are then stimulated, they conduct their information to the central system via thick fibers. In doing so, they can influence the gate so that nociceptive information from the thin fibers is held back and conduction of information from the thick fibers to the central nervous system takes precedence. Using this concept, transcutaneous electrical nerve stimulation (TENS) for pain treatment was developed and implemented. In the same manner, the effectiveness of stimulating the dorsal tracts of the spinal cord using dorsal column stimulation (DCS) is explained. Electrical stimuli should be used to activate thick fibers painlessly so as to prevent the transmission of nociceptive impulses from the thin fibers. We know today that several of these assumptions are not accurate. But the theory triggered important research that transformed our understanding of pain. Further discussions centered on the question of whether noxious stimuli are signaled by specific receptors and guided on neural pathways from the periphery to the brain. Or whether pain is the consequence of a spatiotemporal pattern of interacting inputs from different sensory fibers.

Stimulated by the "gate theory," it was discovered that there is an interaction between the influx from the periphery to the spinal cord as well as higher centers. Noxious stimuli can be modified at the level of the spinal neurons, the brain stem, and higher centers in the brain. Certain mechanisms enhance rather than impede the centripetal passage of nociceptive messages. Both "descending inhibition" and "descending facilitation" have been identified. Descending pain regulatory pathways are subject to feedback inhibition from the periphery as well as cognitive and emotional regulation from brain centers. Intriguingly, there is no absolute, anatomical separation of substrates subserving these processes and the stimulation of a single supraspinal structure may, via contrasting mechanisms, simultaneously trigger both descending inhibition and facilitation.

6.2   Aspects of physiology and pain pathways

6.2.1   Prespinal level

Anatomically, a noxious impulse first activates sensory receptors that are capable of transducing and encoding noxious stimuli. Pain is conducted by thin afferent nerve fibers, which together with the stimulus receptor are called "primary nociceptive afferents." The receptors are the free nerve endings of primary afferent A-δ and C fibers, through which the stimuli are turned into electrical signals, which are then conducted from the prespinal level to the central nervous system. On the way to the brain, the first neuron forms the beginning of the nociceptive pathway; its cell body is located in the spinal ganglion (Figure 1).

The pain fibers react with their receptors selectively to mechanical, chemical, or thermal events. Some are also capable of registering polymodal mechanothermal stimuli and can also be stimulated by chemical irritation. Substances that can trigger burning via chemical irritation are capsaicin or bradykinin, for example, while histamine triggers itching. Thus, specific nociceptors can be distinguished from modality-unspecific polymodal nociceptors by their response to different stimulus modalities. It is still unclear whether they have a differentiated receptor system corresponding to the pressure or tactile receptors (Meissner's tactile or pacinian lamellated corpuscles (Figure 18)). Thus far, no specific morphological structures can be allocated to the functions of nociceptors. In particular, the free nerve endings in the skin are known nociceptors, and since they are easily accessible, they have also been studied extensively. Depending on the intensity of the stimulus, the threshold, and their receptive field, the nociceptors send a wide frequency range of electrical impulse series to the nerve for further transduction.

Nerve fibers with varying conduction properties are available for conducting pain. The less myelinated A-δ fibers conduct at 5–33 m/s, the unmyelinated vegetative fibers at < 2 m/s; these different conduction velocities contribute to the fact that in pain, an initial sharp component can be distinguished from a second, duller component. The second component is then perceived more strongly if there is a selective blockade of the fast-conducting A-δ afferents. This example alone shows the possibility for neuromodulation, in which the loss of information from the A-δ afferent changes the information so greatly that the second information is perceived as more intense pain. The peripheral end of the bipolar nerve cells of the first neuron terminates in the receptor. Adequate stimulation leads to the generation of a membrane potential and the receptor potential is conducted as an action potential or an electrically coded impulse series centrally toward the spinal cord.

A-δ and C fibers terminate in the spinal cords dorsal horn. There the electrical signal is transmitted via synapses and with biochemical mediators to the secondary afferent neuron. Stimulating amino acids as well as neuropeptides function as synaptic transmitter substances and neurotransmitters.

Recently the transient receptor potential vanilloid 1 (TRPV1) channel has generated much interest. TRPV1 is expressed in peripheral and supraspinal localizations and contributes to the descending modulation of nociceptive stimuli. Its activation is associated with thermal hypersensitivity. Under research are TRPV1 antagonists, which can block thermal hypersensitivity to numerous inflammogens and are therefore a possible therapeutic tool. TRPV1 is also called the capsaicin receptor. Upon prolonged exposure to capsaicin, its agonistic effect induces a TRPV1 activity decrease in the peripheral nerve and can thereby induce desensitization (Sect. 10.3.7). This can be helpful in disease states such as painful neuropathy or postherpetic neuralgia.

6.2.2 Spinal level

The afferent fibers terminate in the posterior gray matter or dorsal horn. Neurons in the dorsal horn are arranged in a series of six laminae. From these cells of termination, secondary neurons connect with ventral and lateral horn cells in the same and adjacent spinal segments and subserve both somatic and autonomic reflexes. Other secondary neurons project contralaterally to higher levels. They decussate in the anterior spinal commissure, and ascend via the spinothalamic tract to nuclei in the brain stem and thalamic structures.

The spinal cord actively amplifies the spinal nociceptive processing because nociceptive neurons change their excitability to inputs from the periphery under painful conditions. Sensory fibers terminate in the different laminae of the dorsal horn. A-δ, A-β, and C-fibers connect to matching second-order neurons. Wide-dynamic range neurons (WDR) receive input from all three fibers. WDRs, which are primarily encountered in deeper spinal laminae, encode both innocuous and noxious infor-

mation from the skin and other organs in a stimulus-dependent fashion. Their sensitization by repetitive, nociceptive stimulation plays a key role in the induction of long-term, clinical, inflammatory, and/or neuropathic painful states. The activity of these neurons can be influenced by both excitatory glutamatic and inhibitory GABAergic interneurons. Inhibitory GABAergic spinal interneurons have been shown to undergo apoptosis after nerve injury with increased excitatory transmission. This mechanism increases postinjury pain sensitivity.

A large number of spinal and brain neurons express the neurokinin 1 receptor (NK1) for substance P (SP). SP is the first responder to most noxious and extreme stimuli. SP and its induced cytokines are involved in processes such as pain, inflammation, vasodilation, mood, anxiety, and learning. Part of the antinociceptive effect of noradrenergic-receptor stimulation is thought to be caused by the inhibition of SP release with selective NK1 antagonists. NK1 receptors for SP exist in many spinal neurons. These neurons project to the brain stem and thereby have control on descending inhibition and facilitation. Selective NK1 antagonists are therefore potential candidates for new pain-relieving drugs.

Astrocytes and microglia can also play a role, especially in disease states. Other spinal cord neurons are involved in nociceptive motor reflexes, more complex motor behavior such as avoidance of movements, and the generation of autonomic reflexes that are elicited by noxious stimuli.

6.2.3 Brain level

Two networks have been discovered in the brain. The first is the lateral system, which is responsible for the sensory aspect of pain – that is, the location and duration of pain. The second is the medial pain system, which produces the emotional aspects of pain, the unpleasantness. The conscious pain response is produced by the thalamocortical system.

Anatomically, the lateral system projects through the lateral thalamic nuclei to brain regions including the primary and secondary somatosensory cortices (SI and SII) in the postcentral gyrus. The medial pain system projects through the medi-

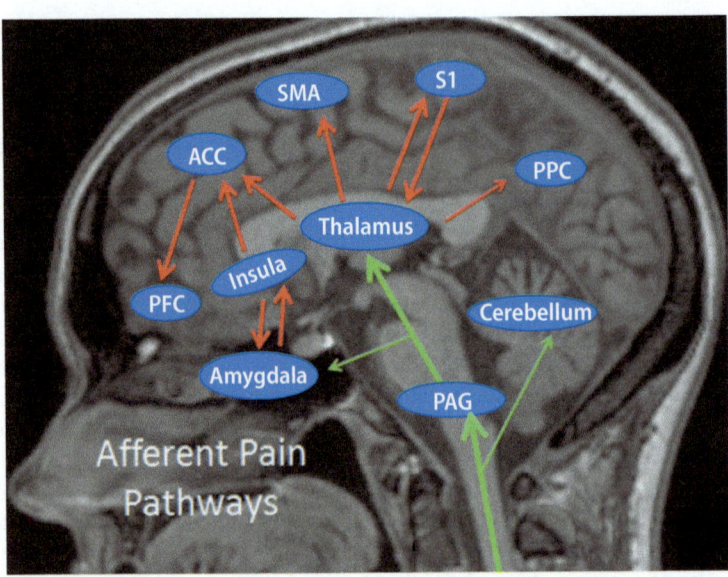

Figure 72 Important brain areas that are activated by the flow of afferent sensory activity. The interplay of the areas forms a response to a noxious signal. *PAG* periaqueductal gray, *PFC* prefrontal cortex, *ACC* anterior cingulate cortex, *SMA* supplementary motor area, *S1* primary somatosensory cortex, *PPC* posterior parietal cortex

al thalamic nuclei. This consists of relay nuclei in the central and medial thalamus, the anterior cingulate cortex (ACC), the insula, and the prefrontal cortex (Figure 72).

Secondary afferent neurons connect via the spinothalamic, spinoreticular, and spinomesencephalic tract to the brain stem and then to the brain. The thalamus is often thought of as the sensory gateway to the brain and, as such, its function plays a crucial role in sensory processing and signal modulation. Within the brain, pain is not processed in a single cortical area, but as an interplay of several distributed brain regions. Functional imaging techniques have revealed the normative network of cortical and brain representations of pain that are consistently activated in response to pain. Also it has been shown that different pathways are involved in acute and chronic pain.

The spinothalamic and spinoreticular tracts deliver nociceptive inputs to higher centers in the brain but also form a loop involving supraspinal structures in the lower brain stem and midbrain. This endogenous modulatory system affects sensations of pain and is known as descending inhibition or diffuse noxious inhibitory controls (DNIC). The output of spinal nociceptive neurons is influenced by descending, propriospinal, and segmental systems, which may locally, diffusely, tonically, or phasically inhibit or facilitate the spinal transmission of nociceptive signals.

The midbrain and the rostral ventromedial medulla are the areas where the periaqueductal gray (PAG) is located. The PAG was the first brain area shown to exert a powerful pain-inhibitory action. Both the PAG and rostral ventromedial medulla (RVM) receive direct projections from the spinal dorsal horn and, thus, they control the ascending nociceptive input by a feedback mechanism.

Descending pathways from the PAG connect via the dorsolateral funiculus to the nociceptive neurons in the dorsal horn and inhibit or facilitate pain transmission. Based on their physiological response properties, spinally projecting RVM neurons can be classified into three types. "On cells" give an excitatory response to a noxious stimulus starting just before a spinal nocifensive reflex. The "off cells" produce an inhibitory response to a noxious stimulus starting just before a spinal nocifensive reflex. The "neutral cells" give variable responses or are unresponsive to noxious stimuli.

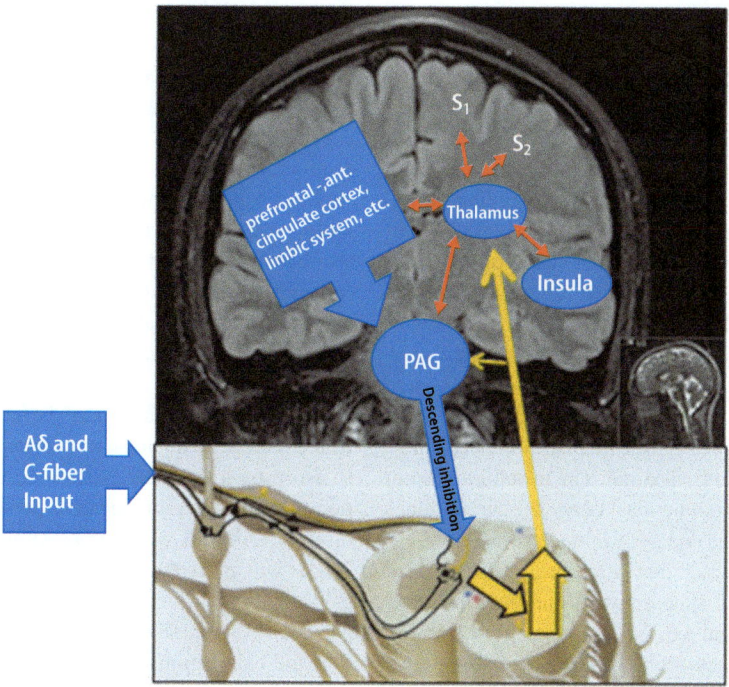

Figure 73 A simplified image showing how afferent sensory information travels from the prespinal level to the spinal cord and crosses to the other side of the cord where it is conveyed to the upper CNS centers. There the information is processed in different areas. Incoming afferent sensory information can be modulated by centrally derived descending inhibition at the spinal cord level. *PAG* periaqueductal gray, *S1* primary somatosensory cortex, *S2*

The midbrain and ventromedial medulla have a high concentration of endogenous opioids. There are also opioid receptors that play a role in the analgesic effect of opioids.

For signal transmission, the descending pathways use the monoaminergic neurotransmitters noradrenaline and serotonin. Norepinephrine reacts in the dorsal horn with α2-adrenoreceptors. Thereby primary afferent terminals are inhibited and the discharge of projection neurons is suppressed. The descending facilitatory action is driven by a serotonergic mechanism, especially in chronic pain states.

The corticospinal pathways inhibit the activity in nociceptive dorsal horn neurons with the transmitter norepinephrine and serotonin. Antidepressants diminish the reuptake of these transmitters into the secreting monoaminergic neuron. The concentration of these substances is thus enhanced extracellularly and they are ready to bind to their receptors on the surface of nociceptive neurons.

A further inhibitory effect on these neurons is imposed by GABAergic interneurons. This effect is lost in peripheral nerve injury when these neurons are under a constant barrage of activity. Owing to glutamate-influenced and glutamate-activated NMDA channels, the neurons are overloaded with Ca ions, which leads to glutamate-induced neurotoxic cell death of these inhibitory interneurons. When these morphological changes are irreversible, a chronic central pain state has been established that is independent of peripheral input and is now a disease of its own. It is a matter of debate whether the extent of this loss of interneurons with the ensuing effect on pain conduction pathways could be one reason for the insufficient or absent response of neuropathic pain toward opioid medication.

The PAG sends descending inputs to the rostroventral medulla, which in turn is controlled by the hypothalamus, the amygdala, and parts of the cerebral cortex (Figure 73). Thereby the brain can control its ascending noxious inputs. It is assumed that cognitive processes can alter the gain of nociceptive signals coming from the spinal cord. This is one explanation for the mechanism of the placebo effect (Schedlowski et al. Pharmacol Rev 2015;67:697–730). It has been shown that the combination of expectation and conditioning can reduce dorsal horn activation and thereby pain perception. Neuroimaging studies of placebo analgesia showed there was decreased activity in the classic pain-processing locations such as the PAG, thalamus, insula, somatosensory cortex, and mid-cingulate area. After painful stimulation, placebo treatment could also substantially reduce pain-related activity in the spinal dorsal horn cells. The placebo effect can be antagonized with naloxone. The μ-opioid receptor antagonist naloxone has been shown to block placebo-induced analgesia in nociceptive pain.

Another factor that influences pain is stress. This emotional state activates the hypothalamic-pituitary-adrenal axis by releasing the corticotropin-releasing factor in the hypothalamus. A potential novel avenue for modulating the experience of pain seems to be oxytocin (Kessner et al. JAMA 2013;310:1733). It is a neuropeptide elaborated by the hypothalamic paraventricular and supraoptic nuclei. It suppresses nociception and promotes analgesia by central and peripheral psychological and physiological processes. Owing to endocrine mechanisms, a modulation of pain takes place and can trigger psychosomatic-induced pain behavior.

Whole-body hyperalgesia and widespread pain are suggestive of structural changes in pain modulatory systems due to psychological factors (stress-induces hyperalgesia). Thus, patients with chronic pain can have biochemical and anatomical alterations or dysfunctions in regions responsible for emotional and cognitive pain modulation. As a consequence, this results in deficits in descending inhibition or activation of descending facilitation and inhibits or facilitates nociceptive processing for the entire body.

Anatomical, physiological, and metabolic knowledge helps us to understand different aspects of pain. But besides this, psychological mechanisms are also part of the complexity of pain perception. Pain expectation alone is already associated with significant activations in the thalamus and the sec-

ond sensory and insular cortex. However, it is usually the affective component that induces a behavioral response. Thereby, pain from a noxious impulse will not always be felt as an unpleasant experience. The context in which a painful stimulus occurs affects how it is perceived. Muscle pain in sports can be motivating. An Olympic champion said: "Pain is temporary, victory is eternal." It has also been shown that witnessing a person in pain can activate pain-related brain regions. The degree of response can be related to the positive or negative cognitive and emotional state of the observer, which can be summarized as empathy. These observations demonstrate that pain can be felt even in states where there is no noxious input from the periphery. It also proves that pain perception can be conditioned by positive or negative personality structures, live events, or other psychological factors. This is why more recently the emphasis in chronic pain is being put on its character. In many chronic pain states the causal relationship between nociception and pain is not tight and pain does not reflect tissue damage. Rather, psychological and social factors seem to influence pain, for example, in many cases of low back pain, fibromyalgia or widespread muscle pain, irritable bowel syndrome, or temporomandibular joint disorder. Studying genetic and environmental influences in large twin studies allows us to discover the general genetic, environmental, and social effects that affect the risk of developing widespread pain (WSP) syndromes. In twin studies by Momi et al. in 2015, using the TwinsUK Registry, it was shown that WSP and neuropathic pain seem to be clinical conditions with common risk factors. These are both environmental and heritable factors, and highlights their relative contribution and interactions. WSP can be accompanied by neuroendocrine dysregulation resulting in fatigue and impaired physical and even mental performance. Possible reasons for such sex-specific predisposal may depend on particular modes of limbic system activation, and also regulatory effects of gonadal hormones on these neural structures. Women are more often afflicted with these chronic pain conditions than are men, which implies that pain perception can be different in men and women. Painful symptoms similar to WSP have been commonly reported in women having premenstrual

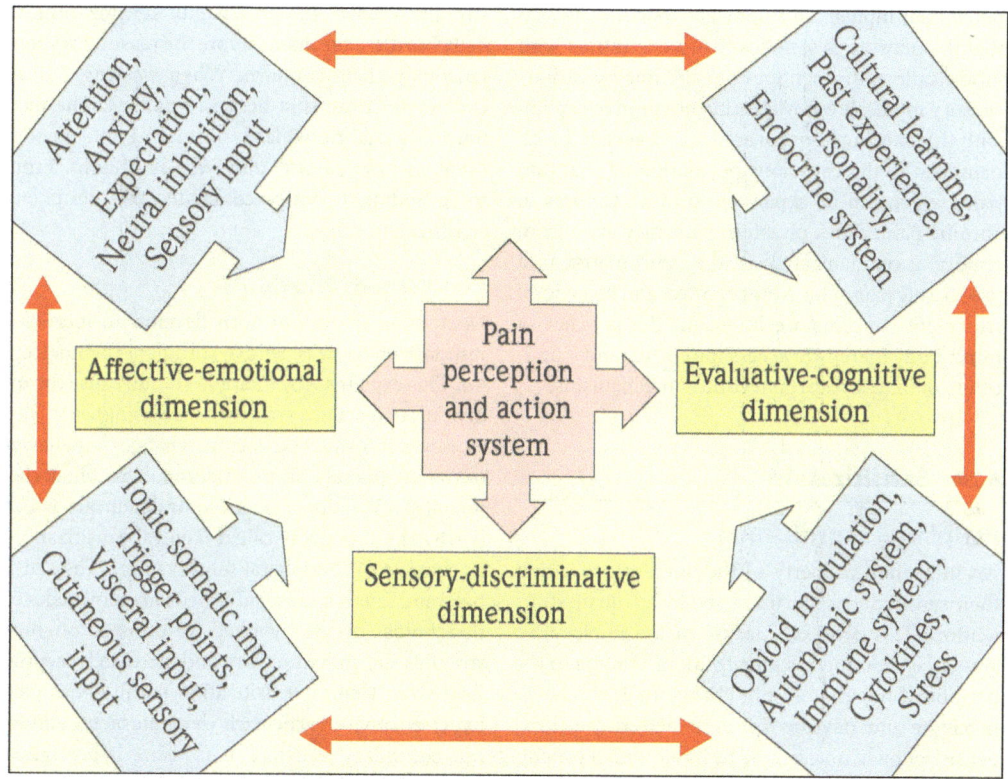

Figure 74 Neuromatrix theory. Multiple features contribute to pain perception and action

syndrome, a cyclic condition that is characterized by headache and various types of musculoskeletal pains, arthralgia, and bowel irritation in the days preceding menses. This suggests a sufficient dependence of these painful symptoms on the changing progesterone/estrogen ratio.

There are disease states in the absence of peripheral changes and inputs that are generally the prerequisite for central sensitization. Nevertheless, within these states, central modulations take place that lead to the experience of pain when confronted with otherwise nonpainful stimuli. New techniques such as functional magnetic resonance imaging and positron emission tomography have shown changes in the brain representing enhanced excitability. In somatoform disorders or disease states like fibromyalgia and widespread pain disorders these changes can be interpreted as a sensitization of supraspinal structures. To date, not much is known about the mechanisms that induce this sensitization.

The supraspinal activity seems to influence the spinal processing of nociceptive input in a top-down fashion.

For this complex interaction, Melzack proposed as early as in 1999 multiple interacting circuits, the neuromatrix theory (Figure 74). In this conceptual model, pain is produced by the output of a widely distributed neural network in the brain rather than directly by sensory input evoked by injury, inflammation, or other pathology. The neuromatrix is modified by sensory experience and is the primary mechanism that generates the neural pattern that produces pain. This embodies an extensive network connecting the thalamus, the cortex, and the limbic system. The concept extends beyond body schema theory and incorporates important psychology theory such as the conscious awareness of oneself. This theory proposes that conscious awareness and the perception of self are generated in the brain via patterns of input that can be modified by different perceptual inputs. An important new idea is also that the network is genetically predetermined with modifications throughout one's lifetime by various sensory inputs. It was Melzack's intention to explain with this gate how certain activities associated with pain lead to the conscious perception of phantom pain, which can be expanded to other features of chronic pain. Thus, phantom pain may arise from abnormal reorganization in the neuromatrix to a preexisting pain state. New cognitive and behavioral treatment concepts are based on the concept of mind-body therapy to access the cortex, and a multitude of intracortical gain control mechanisms.

6.2.4 Sensitization

Peripheral sensitization

An important property of the nociceptors is that their sensitivity can be increased by adequate stimulation. This process, specific to the nociceptive system, is described as sensitization. Trauma to the peripheral nociceptors can change their discharge behavior and develop "peripheral sensitization." Sensitization of nociceptors by inflammatory mediators is induced within a few minutes. Physiologically, it is initially associated with a lowering of the stimulus threshold, with the result that a response is easier to trigger and the stimulus response is enhanced. The increase in activity can be so pronounced that the spectrum of activity can develop to continuous spontaneous activity. Clinically, this is demonstrated by hyperalgesia occurring in the affected area. Stimuli that are normally not painful or only slightly painful are now perceived as severe pain; this mechanism is known as primary hyperalgesia. An example of such a peripheral sensitization with primary hyperalgesia is the increase in sensitivity in thermal nociceptors. Initially, inflammation creates sensitization of the primary nociceptive afferents, which leads to the lowering of their arousal threshold below body temperature. This explains why the increased heat (calor) associated with inflammation can trigger spontaneous pain. This is clinically relevant if a patient shows hyperalgesia to heat stimuli during clinical testing. Such examinations are possible in the context of a quantitative sensory test protocol (QST). QST helps with defined

stimuli to detect the presence of sensitization. In daily life these mechanisms are the reason for spontaneous pain in sunburn. When wearing a shirt, contact with the skin becomes painful. The light touch of the fabric, which is normally a non-noxious event, is now painful; allodynia is present. From these finding, it can be concluded that peripheral sensitization exists.

Central sensitization

Neurons in the dorsal horn develop an increased responsiveness by repetitive stimuli at the nociceptor. This explains why multiple stimuli cause the individual to react to a noxa with increasing sensitivity. However, this situation does not change the behavior of the peripheral and first neuron. This phenomenon of sensitization changes central neurons, which is why the process is called "central sensitization" in contrast to "peripheral sensitization." The sensitized and thus hyperexcitable neurons show reduced thresholds, greater evoked responses, ongoing stimulus-dependent activity, and increased receptive field sizes. Central sensitization manifests as pain hypersensitivity, particularly dynamic tactile allodynia, secondary punctate or pressure hyperalgesia, aftersensations, and enhanced temporal summation. Long-lasting stimuli can lead to an adaptation of the receptor, but in which the subjective perception of pain still continues. This is a process that occurs centrally in the spinal cord and is a wind-up effect. It is explained by a temporal summation of postsynaptic activity in the dorsal horn neurons. The transmission of information between the neurons occurs via synapses and their transmitters.

Neuronal plasticity is achieved by modulating the receptor and channel pattern and incorporating new receptors into the neuronal cell surface. The postsynaptic cells of the dorsal horn such as the interneurons and wide dynamic range (WDR) neurons activate receptors such as amino-3-hydroxy-5-methyl-4-isoxazolepropionic acid (AMPA), the neurokinin receptor, and NMDA receptor complex. AMPA and neurokinins prime the NMDA receptor. Following priming of the NMDA, subsequent glutamate release results in further activation with removal of a magnesium-controlled gate and subsequent calcium influx. In addition, voltage-gated calcium channel activity is upregulated. This mechanism, including the wind-up phenomenon, seems to be the most important central sensitization mechanism. Gabapentin and pregabalin bind to a subunit of the voltage-gated calcium channel, mainly at the central terminal of the primary afferents. This reduces the release of excitatory neurotransmitters and attenuates the changes on the postsynaptic side. In injured primary afferent neurons a downregulation of pre- and postsynaptic μ-opioid receptors can be seen. This loss can lead to an enhanced glutamate release. Later, the central process may take on a momentum of its own and persist in the absence of nociceptive impulses from the periphery.

The phenomena shown here demonstrate that sensitization of the pain-processing system occurs not only peripherally, but also centrally. Thus, a condition of increased arousal is triggered after painful stimulation in the nociceptive system. As a result, stimuli are perceived differently and stimuli that are not painful can be suddenly perceived to be painful. In unfavorable cases, they can sustain pain alone as well. The longer this condition continues, the more probable are changes at the synapses and arousal-transmitting structures, which subsequently change the activation level of the receptors. Although in the normal system, pain is initially perceived only when certain thresholds are exceeded, this is now changed. In the changed system, the pain thresholds are clearly lower and therefore other stimuli can now more easily trigger sensations that are perceived centrally as pain. An important consequence of this knowledge is that pain must be treated promptly and systematically to prevent its chronification. The strategy has to be to desensitize the peripheral and central nervous system.

6.2.5 Pain memory

This wind-up phenomenon is usually short lived. There are also long-lived effects with central sensitization and they can have a profound effect on pain states. Prolonged noxious input can lead to long-term potentiation (LTP) in postsynaptic nerve cells. It is a persistent strengthening of synapses based on recent patterns of activity and a long-lasting strengthening to incoming responses develops. These patterns of synaptic activity produce a long-lasting increase in signal transmission between two neurons. The opposite of LTP is long-term depression, which produces a long-lasting decrease in synaptic strength. LTP is related to learning and therefore the events connected with LTP are believed to form the "pain memory."

When stimulation persists over a longer period, an acute pain state can be transformed into a chronic pain state. This manifestation of use-dependent synaptic plasticity of glutamatergic neurotransmission affects the metabolism and cytoarchitecture of the neurons. Furthermore, glial cells remodel neuronal synapses to intensify nociceptive transmission. These changes can become permanent via altered gene transcriptions. By activation of intracellular signal-transduction cascades, receptors and ion channels are modulated. Studies in neuropathic pain animal models have shown many molecular targets and revealed a high similarity in mechanisms underlying pain amplification and learning and memory in areas of the brain such as the hippocampus and cerebral cortex. This process of neuroplasticity leads from a "peripheral" to a "central" pain. The so-called phantom limb phenomenon illustrates how the mind may retain its ability to experience pain, even after the nociceptors are no longer present. Painful experiences leave a memory trace of pain. If new sensory input activates the memory trace, a gentle touch can be magnified to pain. The memory can also maintain pain by centrally altered cell functions and central sensitization. This sort of pain memory persists and needs no further noxious input from the peripheral nociceptors. Now the pain is chronic and has lost its physiologically important warning function. This is the important transitional step from acute pain to a chronic pain state. It is now a disease in its own right, a disease of the nervous system.

Reversing this pain memory may be one way to end chronic pain disorders. The concept of reconsolidation is important for memory and learning. It is a memory-updating process during which a memory trace can become labile and susceptible to revision with behavioral intervention or pharmacological manipulation. It has to be considered that cognitive and emotional factors have a surprisingly important influence on the upper brain centers where pain perception takes place and influences the adaptive nature of pain plasticity. There are intensive ongoing scientific studies to get an even better understanding of these complex interactions. Only with a better insight into this puzzle of pain we will be able to establish useful and more effective therapeutic interventions. Currently, pharmacological pain management is limited to drugs that are either old or with a limited efficacy. Furthermore, there is active research on ion channels. They are present in peripheral sensory neurons and dorsal root ganglia and are influential in both spinal and supraspinal pain modulation.

6.3   Nociceptors in muscles and joints

The nociceptive afferents in the musculature and joint capsule are similar to those in the skin. There are polymodal free nerve endings belonging to the A-δ group and the C fibers. A-δ fibers are fast-conducting myelinated nerves activated by mechanothermal receptors and high-threshold mechanoreceptors. A-β fibers have a thick myelin sheath and are therefore fast conducting; they register proprioception and touch. Proportionally to their intensity of stimulation, they regulate their mode and frequency of discharge. The majority of nociceptors consist of C-fibers. These nonmyelinated, slow-conducting fibers have a higher threshold than the A-δ and A-β nociceptors. Owing to their polymodality, they respond to thermal and mechanical stimulation but, in addition to A fibers, also to chemical stimuli. This produces the sensation of itching.

  These nociceptors are not activated by the normal activity of the musculature. As in the skin, these nerve endings can, however, be activated by mechanical, chemical, and thermal stimuli. Their polymodality causes them to react differently to various stimuli; noxa can thus be distinguished from one another. Severe mechanical pressure, temperatures over 43°C, or substances such as serotonin, bradykinin, histamine, prostaglandin, and other prostaglandin derivatives such as potassium ions are released as a result of tissue damage and remain in the damaged area for a certain period of time. These processes are relevant in the context of inflammatory changes or injuries to the muscle (Figure 89). There are nociceptors in the joint that do not react to movement or to heavy mechanical strain. But if there are inflammatory changes in this region, the nociceptors are sensitized by the inflammation and then react as sensitive indicators of movement. Nociceptors that require such a special activation are known as sleeping nociceptors. They are the reason that joint inflammation suddenly leads to considerable pain at only the slightest movement. Nociceptors are not only receptive structures. They also have excretory capacity and can release neuropeptides as endogenous substances. This function exists both at the peripheral and central ending. Among these peptides are SP and calcitonin gene-related peptide (CGRP).

6.4   Neurogenic inflammation

When primary sensory afferents are activated, as in migraine, neuropeptides are secreted and can lead to neurogenic inflammation. The term "neurogenic inflammation" refers to an inflammatory state that can give rise to the serious problem of neuropathic pain. It can occur as a consequence of lesions to neuronal tissue but also following noxious events and inflammation in tissue (see also Sect. 10.3.2 "Nonsteroidal anti-inflammatory drugs"). Injuries to neuronal tissue can be toxic as in chemotherapy, metabolic as in diabetes, and viral as with herpes infection. Degenerative changes in bones or disc herniation can stretch, compress, or inflame nerves or nerve roots. Pain-producing substances can be released by activity in the nociceptor itself. For example, the release of CGRP leads to widening in the arterioles and causes warming in the tissue. SP released from nociceptors causes plasma extravasation, in which plasma from the vessels passes into the tissue and creates swelling. The mast cells, which release histamine, are also involved in this process; nitrous oxide (NO) has a mediating effect (Figure 89). In the course of tissue trauma, pain-producing substances leak out from the damaged cells. In addition, these stimuli attract neutrophils and other leukocytes to the site and start the inflammatory response. Their activation together with mast cells is an important step in initiating a cascade of events: the inflammatory process. It results in explosive degranulation and subsequent release of inflammatory mediators into the local microenvironment.

Parts of this inflammatory response are a lowered pH, bradykinin, histamine, platelet-activating factor, adenosine, serotonin, prostaglandins, and leukotrienes. They act on blood vessels and promote plasma extravasation. Plasma extravasation is a critical component in inflammation, because the increase in vascular permeability and subsequent extravasation of plasma that is initiated early in the inflammatory response is a first line of defense required for several other aspects of the inflammatory response, and is the principal determinant of the molecular environment of inflamed tissue. Chronic accumulation of leukocytes at sites of inflammation and production of neutrophil-derived reactive oxygen species contribute to the development of the tissue injury associated with a variety of inflammatory disease states.

Clinically important chemical mediators of primary hyperalgesia include the prostanoids – prostaglandins E2, E1, and I2 (PGE2, PGE1, and PGI2, respectively) – as well as adenosine, serotonin, epinephrine, endothelin, and nerve growth factor. These mediators are thought to act directly, through G protein-coupled receptors or receptor tyrosine kinases, on the peripheral terminals of primary afferent nociceptors. Activation of these receptors triggers specific second-messenger cascades that ultimately alter the excitability of the nerve terminal by phosphorylating ion channels such as the sodium channels. In contrast to the hyperalgesia caused by agents acting directly on nociceptive nerve terminals, agents that act indirectly through other cells in inflamed tissue can also cause acute hyperalgesia. Bradykinin, which is generated from plasma kininogens at the site of an inflammatory response, is one such indirect acting agent. These substances change the sensitivity threshold of the nociceptors, so that non-noxious stimuli are perceived as painful (peripheral sensitization). If this sensitivity is not limited to the original damage area, but extends to adjacent regions, it leads to additional sensitization, which takes place at central structures (see Sect. 1.2 "Sensory system"). All of these processes are initially dynamic, reversible, and last far longer than the short event that caused the damage, so that they also create the bridge to reparative tissue reactions.

6.5   Nociceptive pain

Nociceptive pain is an acute or chronic pain event in which the peripheral and central neuronal structures are intact. This type of pain includes, for example, the pain in inflamed tissue such as in rheumatic arthritis, parts of tumor pain, and back pain or muscle pain. Within chronic musculoskeletal pain, IASP limits nociceptive pain to a disease process which directly affects bones, joints, muscles, or related soft tissue. But when inflammation, autoimmune, or metabolic etiology affects the aforementioned structures and causes lesions in the somatosensory system, the pain is neuropathic. The coincidence of neuropathic and nociceptive pain is the hallmark of mixed pain syndrome (see Sect. 6.7).

Chronic visceral pain originates from the internal organs. When the affected internal structures share sensory innervation with skin, muscle, or subcutis, visceral pain can occur. In situations where the etiology of pain is not well understood, it should be classified as chronic primary pain. Nonspecific back pain, chronic widespread pain, or fibromyalgia falls into this category. The IASP proposes to introduce the new diagnostic entity "primary chronic pain" in recognition of conditions of undefined etiology that affect a broad group of patients who would otherwise be neglected.

It is still a matter of scientific debate how a painful mechanical stimulus is converted into an electrical signal that can then be propagated along sensory nerves to the central nervous system. The exposed nature of sensory "free" nerve endings in joints means that the axolemma of these fibers is probably subjected to significant stretch during joint movement. Recently, mechano-gated ion channels were discovered. The current theory is that movement of the joint generates shear stresses on the axolemma of the "free" nerve endings, resulting in the opening of mechano-gated ion channels. This leads to a depolarization of the nerve terminal and the generation of action potentials, which are subsequently transmitted to the central nervous system where they are decoded into mechanosensation. If a noxious movement is applied to the joint, the firing rate of the afferent nerve increases dramatically and the central nervous system interprets this nociceptive activity as pain. The activation mechanisms that take effect in inflammatory changes were described in the previous chapter and are better understood. In the course of continuous pain, sensitization of peripheral and central nociceptive structures occurs in addition to cellular and metabolic changes. However, the special feature of nociceptive pain is that all changes are reversible with targeted treatment of the pain. The healing of a local process is associated with improvement and healing of the pain.

6.6   Neuropathic pain

The term neuropathic pain embodies a symptom as well as a clinical diagnosis. This pain is generated by a disease-inducing activity within the nociceptive system without adequate stimulation of its peripheral sensory endings. This pain is thus generated by lesions of the somatosensory system (Table 8).

Situations that trigger it exist when there are major mechanical injuries of the body and nerve tissue. However, damage can also be caused by metabolic diseases, as is the case for diabetic neuropathy. Nerve tissue can also be damaged by infections – an example of this is postherpetic neuralgia. This category also includes phantom pain after amputation and pain in paraplegia or lesions of the thalamus as a result of a stroke. For every kind of pain, the patient must first be questioned to determine the exact type. Not every pain connected with a polyneuropathy is neuropathic pain. Myofascial symptoms can occur in the musculoskeletal system from bad posture or movement disorders and thus require a different therapy than neuropathic pain. To understand the pathophysiology of pain, it is important to know that under normal conditions, nociceptors become active only in the presence of noxious stimuli, never spontaneously. Spontaneous impulse generation of the nociceptor or of the first neuron is thus a functional disorder that can be found in neuropathic pain such as trigeminal neuralgia. The function change can thus continue beyond the actual damage and the pain can become an independent disease. Other clinical phenomena of neuropathic pain are listed in Table 9. As a result of injury to peripheral or central nerves, the biological, morphological, and physiological processes and structures are reorganized. In the central nervous system, the projection zones of the nociceptive system can be changed and thus also the processing of impulse inflow from the periphery. As a result, changes occur in pain control that affect the peripheral nociceptive systems from the central nervous system. The descending inhibitory tracts exert such a regulatory control function by releasing serotonin and noradrenaline. Noradrenaline is a key neurotransmitter of the endogenous pain inhibitory system. It acutely inhibits nociceptive transmission (including that mediated by substance P), potentiates opioid analgesia, and underlies part of the antinociceptive effects of the widely prescribed tricyclic antidepressants. Lesions of noradrenergic neurons, however, result in either normal or reduced pain behavior and variable changes in morphine antinociception, undermining the proposed association be-

Table 8 Mechanisms of the development of neuropathic pain

Disease	Possible pathomechanism
Diabetic neuropathy Trigeminal neuralgia	Atrophy and loss of axons, sprouting new axons, ectopic spontaneous activity, and cross-talk in the peripheral nerve
Entrapment syndrome	Loss of afferent inhibition in a compression-related selective A-δ fiber lesion
Root avulsion	Spontaneous hyperactivity from deafferentation of dorsal horn neurons
Postherpetic neuralgia	Ectopic spontaneous activity in the spinal ganglion, cross-talk, receptor sensitization
Sudeck's syndrome causalgia	Ectopic spontaneous activity in neuromas, α-adrenergic chemosensitivity, receptor sensitization
Root lesions	Mechanosensory spinal ganglia and dorsal roots
Tinel's sign	Mechanosensory ectopic pacemaker in the area of the regenerating axon membrane

Table 9 Terminology of positive and negative sensory signs which can occur in different combinations in neuropathic pain (Source: Baron et al. 2006)

	Symptom	Definition	Test	Response
Negative Signs	Hypoesthesia	Reduced perception of a non-painful touch	Stroking the skin with a cotton swab	Numbness, reduced sensitivity to touch
	Pallhypoesthesia	Reduced perception of vibration	Stimulation of a bone prominence with a tuning fork	Reduced perception of vibration
	Hypoalgesia	Reduced perception of a painful stimulus	Pin-prick testing on the skin	Numbness
Positive Signs Spontaneous Pain	Paraesthesia	Pins and needles-like sensation, tingling	Measure area and define intensity by visual analogue scale (VAS)	
	Lancinating pain attacks	Electric shock-like episodes	Define number of shocks per time, eliciting factors, intensity by VAS	
	Pain localized to the surface	Pain sensation of burning character	Measure area, intensity by VAS	
Positive Signs Evoked Pain	Mechanical dynamic allodynia	Pain perception after a non-painful light touch	Stroking the skin with a cotton swab	Burning, stinging pain
	Mechanical static allodynia	Pain perception after a non-painful light touch	Light pressure with the finger on the skin	Dull pain
	Pin-prick allodynia	Pain perception after a non-painful light touch	Touching the skin with von Frey hair or sharp toothpick	Burning, stinging pain
	Somatic allodynia	Pain perception after a non-painful stimulus	Light pressure on joint or muscle	
	Cold allodynia	Pain perception after a non-painful cold stimulus	Touching the skin with a cold (10 °C) item	Burning, stinging pain
	Heat allodynia	Pain perception after a non-painful warm stimulus	Touching the skin with a warm (45 °C) item	Burning, stinging pain
	Wind-up	Repeated identical pain gets more painful over time	Repeated skin stimulus with a toothpick < 3 s for 30 s	Stinging pain with incremental intensity
Sympathetic Influence	Sympathetic maintained pain	Nerve block leads to long-lasting pain relief	Sympathetic nerve block	Pain relief
	Sympathetic independent pain	Block does not lead to long-lasting pain relief		No pain relief

tween noradrenaline (NA) deficiency and chronic pain (hyperalgesia). Neurotransmitters change the phenomenon of pain so that the sensory, affective-motivational, somatic, and vegetative as well as motor components of experiencing pain can be altered. Chronic neuropathic pain is characterized by the continued existence of pain after the injury has healed. After the tissue is healed, the pain may even increase in intensity and extent. In contrast to nociceptive pain, it is much more difficult to treat neuropathic pain. Clinical symptoms of neuropathic pain and the associated pathomechanisms are listed here and the appropriate examination method is described.

Spontaneous pain

This pain can occur as a permanent burning sensation or in the form of sharp attacks. It is the most frequently occurring form of neuropathic pain. The permanent burning pain is located at the surface. The sharp attacks are frequent and of very short duration. With respect to development, it is assumed that there is chronic pathological resting activity in afferent nerves. There are partially damaged C-nociceptors in the skin, in which ectopic stimulation and sensitization occur.

Paresthesia and dysesthesia

These sensory phenomena are frequently described by patients with polyneuropathies. With respect to their etiology, it is assumed that there is pathological spontaneous activity in low-threshold, non-nociceptive A-β afferents.

Heat-induced hyperalgesia

Touching the skin with a warm device (test tube filled with hot water) that is normally not painful elicits pain. This phenomenon has already been briefly explained in Sect. 6.2.4. Heat can induce pain or intensify pain in the innervation area of a damaged nerve. It is the result of the sensitization of a partially damaged C-nociceptor.

Cold-induced allodynia

Touching the skin with a cold device (test tube filled with ice-water) that is normally not painful induces pain. This can be found in some polyneuropathies or as a sequela of a traumatic peripheral nerve lesion. In this situation A-δ fibers are activated and their activity is registered by sensitized second-order dorsal horn neurons (Table 9).

Dynamic allodynia

Gentle light stroking of the skin with a cotton swab induces pain (Table 9).

6.7   Mixed pain concept

The dualistic classification into nociceptor pain or neuropathic pain is theoretical. For teaching purposes, the dual presentation is certainly helpful, but the reality is that the two types of pain overlap and are involved to varying degrees in many pain situations. For this reason it is important to study the respective symptoms to determine the extent to which nociceptive and neuropathic, and thus peripheral and/or central pain components, are present, either alone or in parallel (Tables 9, 10). Corresponding to the situation, targeted therapy can be planned and initiated. This idea, that two theoretically different pain concepts with their respective pathophysiological changes can be involved in the pain situation, was the basis for the development of the unifying "mixed pain" concept (Table 11). The concept becomes clinically significant in particular for degenerative changes in the skeletal system and for back pain syndrome, sciatica, or lumbago as well as for tumor pain. In degenerative spine and joint pain, afferents from joints, ligaments, and muscle tissue – and for tumors, from tumor tissue – are conducted toward the central nervous system via nociceptors, and in addition, neural structures are directly damaged from infiltration or mechanical compression. The mixed pain concept is of special interest for chronic low back pain or sciatica, as the pain states have a high prevalence. In these pain states, nociceptive and neuropathic pain components can be distinguished and they occur in parallel. As explained in Sect. 4.2, there is afferent innervation of the intervertebral disc in healthy subjects only in the outer third of the annulus fibrosus and nociceptive A-δ and C fibers are involved. The situation changes in degenerative disc disease and low back pain when nerve fibers sprout deep into the degenerative intervertebral disc along with blood vessels. In this situation, macrophages from the vessels attack damaged cartilage and a massive release of inflammatory mediators causes chemical damage to nociceptive nerve fibers in the intervertebral disc tissue. Thus, an inflammatory process together with mechanical irritation of the nerves will trigger a nociceptive input. The degenerated disc causes mechanical instability in this segment. The newly sprouted nociceptive nerve fibers are exposed to

Table 10 Clinical phenomena that occur in isolation or in combination in neuropathic pain

- The pain occurs without any recognizable tissue damage

- Abnormal, unusually unpleasant burning or shock-like painful sensations

- Pain occurs with delay after a damaging event

- The area with reduced tactile sensation is painful

- The pain has a paroxysmal sharp or stabbing character

- Even light touching is painful (allodynia)

- Pronounced summation effect and afterpain following repetitive stimuli

mechanical irritation. This suggests that a neuropathic pain component may be involved in patients with no radicular irradiating pain but with diffuse overuse-related back pain, at deep sites in many cases, even in the absence of mechanical compression of the spinal nerve. This combination of sustained stimulation of nociceptive and neuropathic pain mechanisms can be the reason for chronic back pain. In chronic back pain there are, in addition to problems in the disc, concomitant strains on muscle tissue and facets too. Sensitization of muscle and joint nociceptors create an additional nociceptive pain component.

Table 11 Mixed pain: Many pain states are composed of a variable mixture of nociceptive and neuropathic pain mechanismus. This understanding of pain is embedded in the mixed pain concept

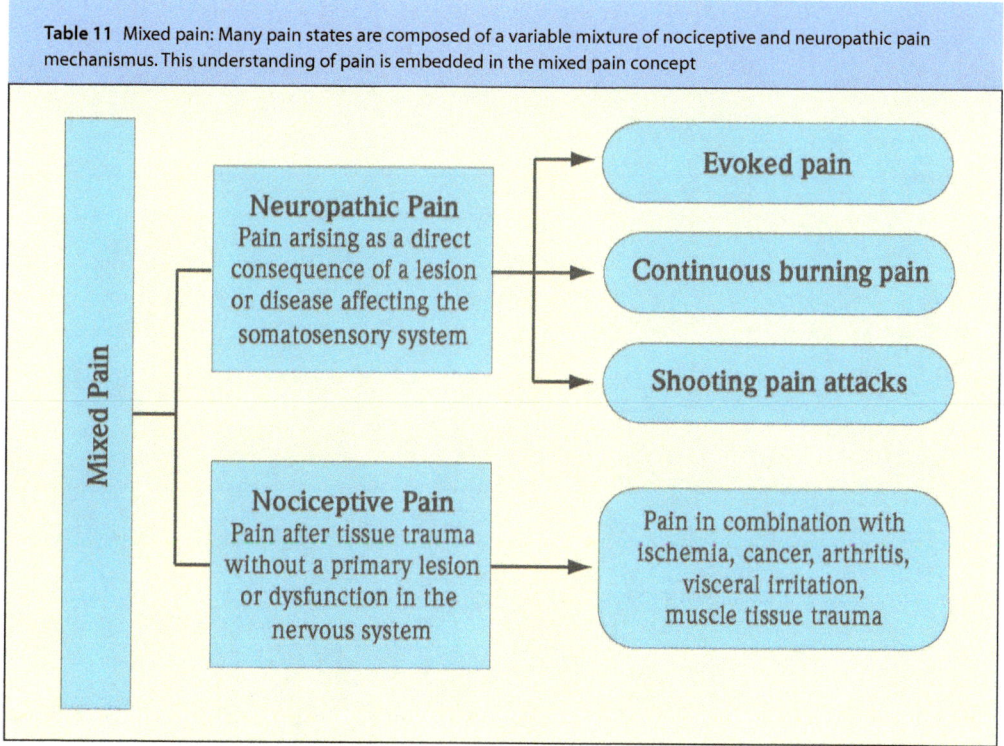

Pain mechanics

The term "peripheral neuropathy" includes, irrespective of etiological aspects, all diseases of the peripheral nerves. Regarding the etiology, peripheral lesions develop from both mechanical and inflammatory damage as well as from metabolic or toxic damage. Thus far, lesions of individual peripheral nerves have been discussed with respect to their motor, sensory, and vegetative disorders. Using these disorders, it is possible to deduce a peripheral nerve structure, lesions, and the location of the lesions in a systematic examination process. This process is used in clinical routine to reach a diagnosis. When evaluating a neuropathic pain state, the aim should be to find out if the pain is based on a lesion in the central or peripheral somatosensory system.

Disease states are often complex, when mainly pain dominates the clinical picture. The location of the lesion, which can be peripheral or central, cannot be directly determined from the patients' pain complaint. In the context of inflammatory, metabolic, or toxic disorders, single focal segments of the nervous system are only rarely changed; more often several structures are regionally damaged as in herpes zoster, or there is a generalized disorder as in polyneuropathy or polyradiculitis. Damage to different tissues such as nerves, muscles, and connective tissue also plays a role. The very important and clinically most frequent disorders are sciatica, back pain, and myofascial pain.

It is not rare to see patients with all the typical clinical signs of radicular compression and a typical radicular radiation pattern, but no proven mechanical impingement of the nerve root and hence no mechanically induced nerve lesion in the true sense.

Animal experiments have shown that the cause of sciatica (see Sect. 9.5.4) can be due to inflammation of the nerve root through anatomical proximity to an inflamed and degenerated intervertebral disc. In this case evidence speaks for an inflammatory neuropathic radicular pain concept. In favor of this hypothesis are elevated concentrations of phospholipase 2 (PLA2). This potent inflammatory mediator was detected in sequestered human intervertebral discs. A positive Lasègue's sign and the severity of the sciatica component correlated with elevated PLA2 levels. In animal studies, PLA2 was injected into the epidural space at level L4–L5 and the effect on the correlating nerve roots that under-

went demyelination was studied. The result of mild mechanical compression of these spinal nerves was ectopic activity that persisted after the mechanical stimulus was released. The animals displayed hypersensitivity to sensory stimuli and a reduction in strength consistent with segmental innervation.

Neuropathic pain in sciatica and low back pain can also be due to aberrant somatosensory processing and goes beyond the normal plasticity of nerve fibers in the intervertebral disc tissue. Thus, an inflammatory process together with mechanical irritation of the nerves will trigger a nociceptive input. The degenerated disc causes mechanical instability in this segment. The newly sprouted nociceptive nerve fibers are exposed to mechanical irritation. This suggests that a neuropathic pain component may be involved in patients with no radicular radiating pain but with diffuse overuse-related back pain, at deep sites in many cases, even in the absence of mechanical compression of the spinal nerve. This combination of sustained stimulation of nociceptive and neuropathic pain mechanisms can be the reason for chronic back pain. In chronic back pain there are, in addition to problems in the disc, concomitant strains on muscle tissue and facets as well. As depicted in Figure 89, the release of neuropeptides from nociceptive neurons can cause local swelling in the muscle and trigger the release of bradykinin and other inflammatory substances, which sensitize muscle and joint nociceptors and create an additional nociceptive pain component.

This knowledge of the combination of acute nociceptive and neuropathic components, which can develop into a chronic form of pain with central sensitization phenomena, is important for daily clinical diagnostics and therapy. The neuropathic pain component is often a significant factor in low back pain. Therefore, important mechanisms for the development of chronic pain, which are of relevance for frequent clinical symptoms, are taken into consideration in the mixed pain concept that is described in Sect. 6.7.

Following, some pathophysiological pain mechanisms will be explained. Their knowledge is helpful for understanding how lesions of the peripheral nerve can contribute to pain and knowing which therapeutic interventions may be helpful.

7.1 First and second neuron pathways

The physiological situation of the peripheral and central pathway for nociceptive afferent impulse conduction is depicted in Figures 73 and 75. Owing to the type of stimulus, afferent information is conveyed to the spinal cord via A-β, A-δ, or C fibers. There, polysynaptic transmission takes place onto the secondary neuron to conduct information within the central nervous system. The primary afferent neuron and receptor are silent when at rest, and respond with activity only when adequate stimulation occurs. The different pathways are morphologically and functionally separated from each other.

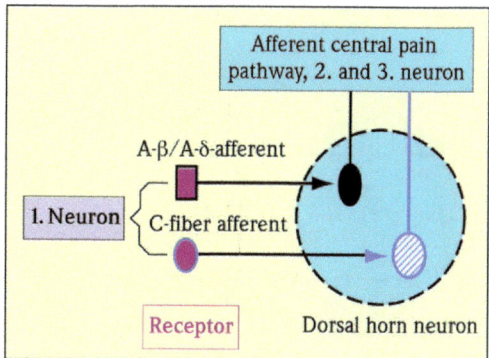

Figure 75 The physiological situation of afferent impulse conduction from the periphery to central areas

7.2 Impulse propagation in the normal nerve: the diseased nerve and cross-talk

The peripheral nerve fibers can be damaged owing to various noxious disease states. This leads to morphological changes on the axon, the myelin sheath, and ion channels. All of the conduction properties of the nerve are influenced and impaired. Figure 76 shows examples of variations of the multiple possible impairments. The upper trace in example A shows a normal nerve. If a given impulse pattern is imposed onto the nerve, the nerve will conduct this pattern undistorted. The lower trace illustrates a diseased state. Owing to altered conduction properties, the generated impulse pattern on the left site reaches the central site on the right in a distorted manner. In B none of the originally coded information is conveyed undistorted to the central nervous system. Frequencies as well as amplitudes are altered and explain how non-noxious information can be falsified by altered conduction properties into noxious information. Nerve damage can distort impulse-encoded information by an additional mechanism. In this instance, the damage allows the unhindered spread of discharge activity from one axon to axons in the neighborhood. This situation is called cross-talk via ephaptic transmission and is indicative of a lesion in neuronal tissue (Figure 78). The ephapse is the anatomical area of damaged nerve fibers, where unphysiological transmission of

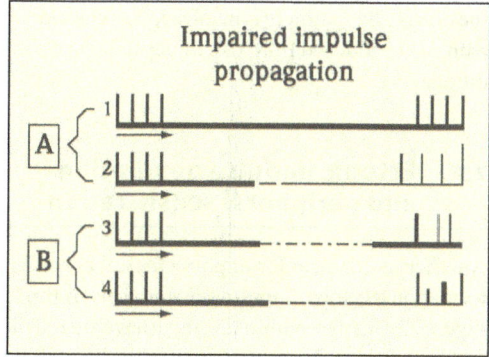

Figure 76 Impulse conduction in normal and damaged nerves

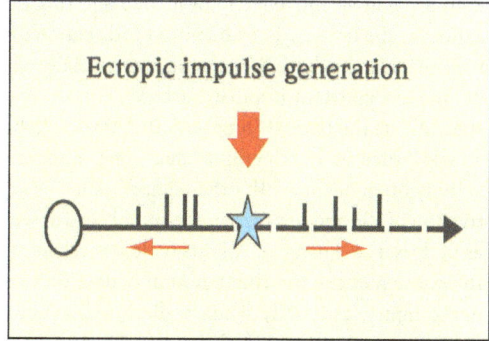

Figure 77 Pain due to ectopic impulse generation. This mechanism is important in peripheral nerve injury as it is often linked with abnormal ectopic electrophysiological activity in nociceptive nerve fibers, cellular activation of glia and neurons in dorsal root ganglia, and the spinal cord neurons. Development of ectopic impulse generation

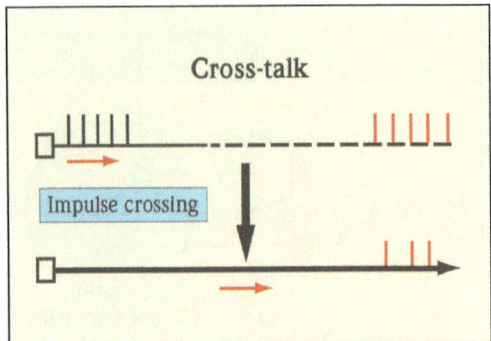

Figure 78 Cross-talk is a pathological ephaptic electrical interaction between fibers

nerve activity between nerve fibers takes place. If the nociceptive pathway is involved, the sensation of pain can be induced from this unphysiological condition.

7.3   Ectopic impulse generation and peripheral sensitization

Another mechanism for unphysiological coded information is ectopic impulse generation as shown in Figure 77. Changes within the nerve have altered the resting and membrane potential, thereby sodium channel dysfunction and de novo synthesis of new receptors and channels play an important role. This destabilization can induce spontaneous activity within the nociceptive receptor as well as along the axon. As the activity is transmitted bidirectionally to both ends, the dorsal horn neurons especially will be under a constant impulse barrage. This disease state affects the patient in episodes of abrupt "pain attacks" such as in trigeminal neuralgia. Ephaptic activity often occurs with other changes such as activation of glia and neurons in the spinal cord as well as in dorsal root ganglia. All this is closely related to the consequences and the temporal course of local nerve injury, especially when wallerian degeneration is present in the case of axotomy. During gradual axoplasmic disintegration, the axolemma fragments and the contents undergo granular dissolution. Axonal breakdown is mediated by calcium influx and involves activation of axonal proteases. A cascade of biochemical and TNF-α-driven proin-

flammatory cytokine processes occurs. TNF-α is widely considered the prototypic proinflammatory cytokine owing to its principal role in initiating the cascade of activation of other cytokines and growth factors in the inflammatory response. Other proinflammatory cytokines, such as interleukin (IL)-1β and IL-δ, are also involved. Cytokines are regulatory proteins that are secreted by white blood cells and a variety of other cells in the body including neurons and glia. After axotomy, macrophages infiltrate the nerve, they secrete components of the complement cascade, coagulation factors, proteases, hydrolases, and also interferons and cytokines. Substance P is also released from nociceptive sensory neurons. T cells released from macrophages are activated and proinflammatory cytokines are produced. Schwann cells (Figure 7) that are not in contact with axons express nerve growth factor (NGF) and an NGF receptor on their surfaces. NGF and the NGF receptor also assist in guiding axons down remnant basal lamina tubes. Excessive NGF signaling can initiate and sustain neuropathic pain. Axotomy deprives the dorsal root ganglia (DRG) of peripheral input. Electrophysiological changes are instantaneous. The area of injury produces a barrage of impulses that immediately alter DRG and spinal cord gene expression by upregulating messenger proteins such as a "transcription cocktail" including c-fos and c-jun. In addition to these electrical and biochemical changes, another form of communication within the axon is affected. Via the retrograde axonal flow (Figure 3), further information is transmitted to the DRG or centrally for the cell nucleus. These effects develop more slowly and have a protracted effect on the function of central neurons. In partially denervated nerves, the biochemical and immunological changes of the damaged nerves communicate with intact fibers that pass within the same nerve. Increased levels of cytokine IL-1 can induce, as in the axotomized neuron, the expression of NGF, and NGF may sensitize nociceptors.

Within the affected neurons, the mode of synthesis and transport of transducer molecules and channel molecules is changed, especially for sodium channels, and new receptors for glutamate and α-adrenoreceptors are expressed. These changes are depicted in Figure 79. Neuron number 0 gives a schematic picture of the normal situation. In num-

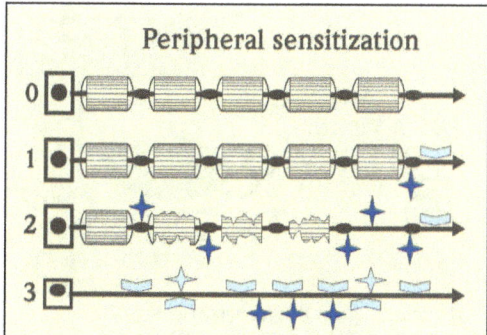

Figure 79 Pain due to peripheral sensitization

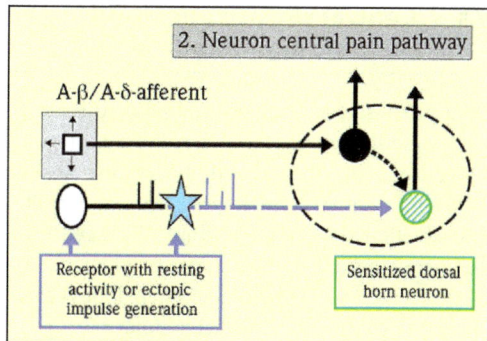

Figure 80 Sensory afferent activity from the periphery induces changes in the dorsal horn neurons

ber 2, destruction of the myelin takes place and a new receptor and channel are expressed. Neuron 1 is undamaged. But owing to the proximity to damaged neuron 2, the nerve is affected and also expresses new receptors and channels such as sodium and calcium channels. The more a neuron is damaged – for example, neuron 3 – the more changes take place, increasing the likelihood of spontaneous activity and cross-talk. All changes described have the effect of peripheral sensitization of the neuron. Nociceptor sensitization is related to changes in voltage-gated sodium channels. Sodium-channel function is of great scientific interest, as drugs affecting these channels have been proven to effectively relieve neuropathic pain. Carbamazepine stabilizes sodium channels in an inactive state and can, for example, suppress painful afferent activity in trigeminal neuralgia. A similar effect can be achieved with other anticonvulsants such as phenytoin or local anesthetics.

7.4 Central sensitization and anatomical plasticity

The mechanisms described thus far are located mainly in the periphery. In neuropathic pain there are, usually as a consequence of peripheral changes, also changes in the central nervous system, which initially affect the polysynaptic connection in the dorsal horn and later other neurons in the brain stem, brain, and cortex. Impulse barrage from the primary afferent neuron can induce sensitization of

the second afferent neuron in the dorsal horn. This central sensitization refers to the augmented response of central signaling neurons. The physiology of the dorsal horn neurons is changed in which spontaneous activity is induced or the amount of already present activity is enhanced. The excitation threshold for incoming mechanical and thermal activity is reduced and the response to such stimuli is prolonged and intensified. The receptive field in the periphery is enlarged. In this state of sensitization the neurons now respond to mechanical and thermal stimuli (A-β and A-δ afferents) toward which they were previously unresponsive. The central changes explain the primary hyperalgesia at the site of tissue injury and the secondary hyperalgesia that occurs in the uninjured tissue surrounding the site of injury. This process is reversible. If nociceptive input induced by inflammation is blocked in the injured zone, secondary hyperalgesia abates promptly.

The situation depicted in Figures 73 and 80 is in reality far more complex. The important message of this scheme is that under C fiber denervation, growth factors that induce nerve growth can be expressed. A-β or A-δ fibers can start to sprout and will establish functional and even morphological connections with dorsal horn neurons that have been sensitized because of damage of the C fibers. This aberrant innervation pattern can establish firm wiring that then makes allodynia a permanent problem.

7.5 Central sensitization and descending inhibition

Pain processing is far more complicated than has been described to date. The dorsal horn integrates ascending, intraspinal, and descending signals. Wind-up and LTP of nociceptive neurons as well as transduction in the spinal cord are influenced by the balance of facilitatory and inhibitory interactions of higher centers in the brain. Brain stem loops govern descending inhibition and facilitation.

Figure 81 presents a simplified schematic drawing of central sensitization and the influence of upper brain centers. A receptor with resting activity sends a barrage of discharge activity to the 2nd neuron in the dorsal horn. There, wind-up takes place and input from peripheral noxa is enhanced centrally; this is primary hyperalgesia. Not only nociceptive neurons in the dorsal horn, but also WDR neurons undergo wind-up and are now easier to be activated by other incoming signals. Before, the WDR neuron did not respond to activity coming from A-β and A-δ afferents. Now the situation has changed and activity from A-β and A-δ afferents activate normally ineffective synapses and the WDR neurons respond. They send impulses centripetally and activate pain-sensitive circuits in the brain (see Figure 72). A nonharmful event such as a gentle stroke of the skin with a cotton swab will now be felt as painful: Allodynia is established.

Higher brain centers (as described in Sect. 6.2.3) can impose a facilitatory or inhibitory control on spinal neurons via descending inhibition or diffuse noxious inhibitory controls (DNIC). Spinal mechanisms mediate the descending pain inhibitory effect by direct (postsynaptic) inhibition or facilitation of spinal pain-relay neurons (Figure 73) or by indirect inhibition of spinal pain-relay neurons through activation of inhibitory interneurons. DNIC also refers to the mechanism by which dorsal horn WDR neurons responsive to stimulation from one location of the body may be inhibited by noxious stimuli (such as heat, high pressure or electric stimulation) applied to another, remote location in the body. This can cause a counter-irritation phenomenon. DNIC may therefore be a mechanism to explain the antinociceptive effect of acupuncture. The inhibition is thought to originate in the brain, and is

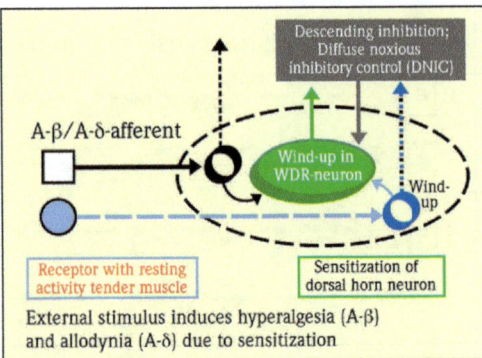

A-β/A-δ-afferent

Descending inhibition; Diffuse noxious inhibitory control (DNIC)

Wind-up in WDR-neuron

Wind-up

Receptor with resting activity tender muscle

Sensitization of dorsal horn neuron

External stimulus induces hyperalgesia (A-β) and allodynia (A-δ) due to sensitization

Figure 81 Central sensitization and descending inhibition

believed to affect both WDR and nociception-specific neurons in the dorsal horn. This descending inhibition can selectively, with different regions of the brain, inhibit certain inputs to the spinal cord.

7.6 Sympathetic-afferent coupling

In normal physiology the sympathetic nervous system has no influence on the primary nociceptive afferent. This can be different for a peripheral nerve lesion. As depicted in Figure 79, profound changes take place on the surface of injured and uninjured nerves. Wallerian degeneration creates a milieu of cytokines and growth factors, and via messenger proteins it induces the expression of new ion channels and receptors, among which are adrenoreceptors. When stimulated, the sympathetic efferents release noradrenalin. There are now theoretically several ways by which this catecholamine can reach the α-receptor to induce nociceptive activity in 1st neuron afferents. Noradrenalin released into the systemic blood circulation can be carried to catecholamine-sensitized axons, bind to the receptor, and elicit nociceptive activity. Another possibility is a close contact between sympathetic nerve fibers and nociceptive afferent fibers. Postganglionic sympathetic axons are known to have a powerful regenerative capacity. After a nerve lesion, the axons can sprout along the sensory nerve and reach the spinal ganglion. There synaptic contacts can be established between sympathetic and nociceptive nerve fibers. Then noradrenalin can bind on α-adrenergic recep-

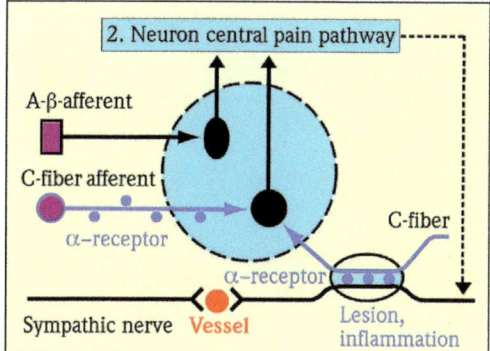

Figure 82 Sympathetic-afferent coupling

tors and establish a sympatho-adrenergic coupling. An as-yet unproven but possible mechanism is that of cross-talk between the different fiber groups via ephaptic impulse transmission. As in the nociceptive system, there is also a central regulatory influence on the activity of the peripheral sympathetic nervous system.

As can be seen in Figure 82, sympathetic fibers have a direct regulatory effect on blood vessels. Owing to sympathetic dysregulation or neurogenic inflammation, the affected tissue suffers vasoconstriction, and blood flow as well as nutrition are impaired. The ensuing effect in the tissue milieu can be hypoxia and acidosis or a reduced buffer capacity. Protons are known to act as important pain mediators that can also induce allodynia. Sympathetic activity can also induce vasodilatation. This situation may allow for plasma extravasation that can, as described, facilitate neurogenic inflammation and activation of nociceptive activity in primary afferent nociceptors.

Polyneuropathies

Mononeuropathies are usually the result of circumscribed traumatic, ischemic, or inflammatory processes affecting individual nerves. Polyneuropathies, on the other hand, affect all peripheral nerves. This systemic damage occurs in connection with generalized inflammatory, infectious, toxic, or immunological problems. Between the two is mononeuritis or mononeuropathy multiplex, in which an underlying systemic disease leads to clinical damage of individual peripheral nerves or nerve roots.

Polyneuropathies have very different causes and a mixture of clinical symptoms. This variation arises partly from the structure of the peripheral nerve with its motor, sensory, and vegetative fibers, which can be affected in isolation or in any combination. Additionally, the location of damage plays a role,

and determines the distribution of the neurological disorders in the body. Morphologicaly, three groups can be differentiated as shown in Figure 83.

The most frequent clinical deficit pattern has the following characteristics: sensitivity disorders (vibration and tactile perception and sense of position, less often affected are sense of pain and temperature) belong to the symmetrical stocking distribution pattern and are distally pronounced. The proprioceptive reflexes are reduced or absent, muscle paresis, usually distal, is the major motor aspect, vegetative deficits occur in the form of sweat secretion disorders. The thin nerve fibers are affected to a greater extent and patients complain of painful paresthesia (dysesthesia), tingling, and temperature dysesthesia.

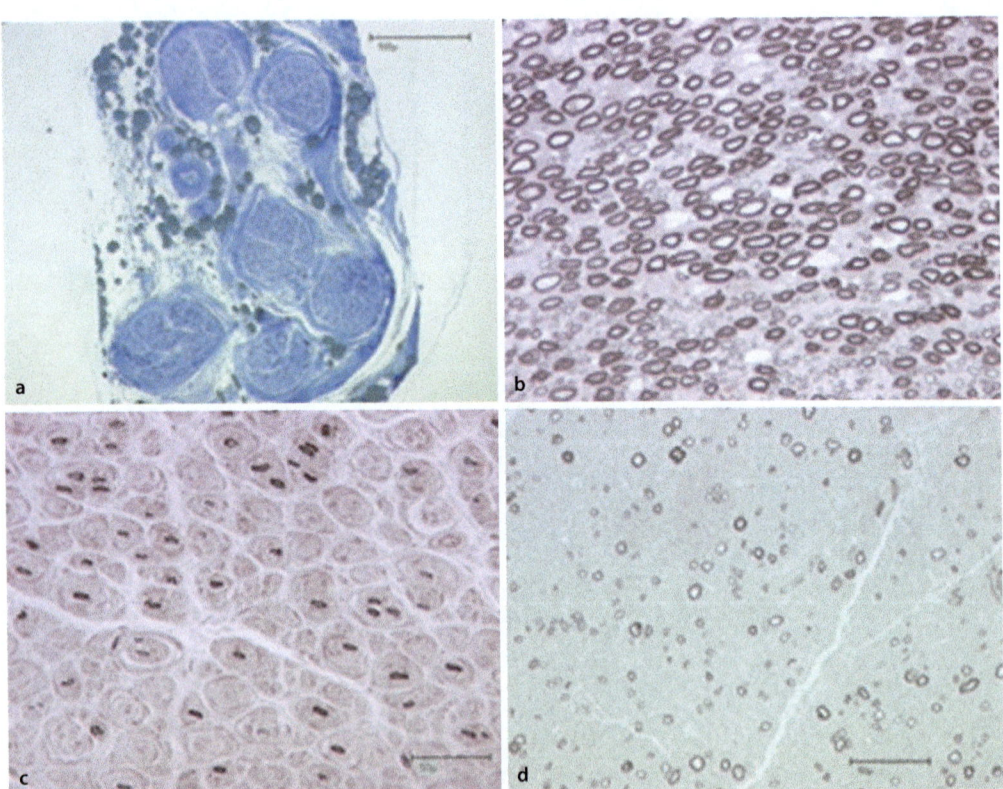

Figure 83 a Normal sural nerve with six fascicles (see Figure 4). **b** Cross section of a normal sural nerve from an adult. **c** Cross section of the sural nerve from an adult. Severe loss of myelinated nerve fibers. Onion-bulb- like lamella of proliferating Schwann cells are wrapped around each small nerve fiber. Chronic demyelinating neuropathy. **d** Cross section of the sural nerve from an adult. Severe reduction of fiber density: Some groups of small myelinated nerve fibers represent regenerating fibers. Chronic neuropathy of axonal type

Table 12 Classification of polyneuropathies

I. Symmetrical polyneuropathy
A Primary distal and sensory polyneuropathy
  Affected: a) mainly large fibers
     b) all fiber types
     c) small fibers
B Chronic proximal polyneuropathy
C Autonomous neuropathy
 	
II. Focal and multifocal neuropathy
A Acute or subacute proximal motor neuropathy
B Cranial nerve neuropathy
C Extremities or trunk mononeuropathy

Polyneuropathies can be classified according to clinical symptoms, etiological aspects, or types of nerve lesions. A classification according to the clinical deficit patterns, acuity of the disease, and fibers involved is shown in Table 12.

Table 13 presents a classification that takes the etiology of frequent polyneuropathies into consideration and categorizes the primary attack site of the noxa on the neuron by axonal degeneration and segmental demyelination.

For diagnostic purposes, a biopsy specimen from the sural nerve is usually obtained. Depending on the level of the biopsy, the nerve, as all other nerves, can have a different number of fascicles. The biopsy can help to differentiate between different types of neuropathy. The pictures are quite typical for hereditary neuropathies when onion bulb formations are present and in pure axonal damage. In chronic disorders regenerating fibers are usually present to some degree (Figure 83).

8.1 Demyelinating neuropathies

In demyelinating neuropathies, the disease process occurs mainly at the Schwann cells and the myelin formed there. Since only the myelin is initially affected, the axon remains undamaged. To restore a demyelinated lesion, it is therefore initially necessary only that the myelin is formed again. The damage first impairs only the thickness of the myelin sheath and the size of the internodal segments and thus NCV. The axon of the nerves remains intact, ensuring trophic supply of the muscle. Therefore, no denerva-

Table 13 Classification of polyneuropathies by etiology and pathophysiology

Axonal degeneration	Segmental demyelination
Metabolic neuropathy	
Hypo/hyperglycemia Uremia Alcoholism early stage Vitamin B_1, B_6 deficiency	Diabetes mellitus Hypothyreosis Alcoholism late stage Vitamin B_{12} deficiency Hepatopathies
Toxic neuropathy	
Arsenic, thalidomide, acrylamide Lead, thallium, vincristine, Cisplatin, phenytoin, Ethambutol, Nitrofurantoin	Amiodarone Chloroquine
Hereditary neuropathies	
Charcot-Marie-Tooth type 2 (neuronal form) Hereditary neuropathies (HMSN II+V) (HSN I–IV)	Charcot-Marie-Tooth type 1,3 (hypertrophic form) Dejerine-Sottas Refsum's disease
Immunologically transmitted neuropathy	
Collagenosis Vasculitis Guillain-Barré syndrome (AMAN)	Acute/chronic/focal inflammatory demyelinating PNP, borreliosis Guillain-Barré syndrome (AIDP) (CIDP) Borreliosis HIV Infections Diphtheria
Dysproteinemic neuropathy	
Myeloma	Lymphoma Gammopathies
Systemic diseases	
Sarcoidosis Porphyria	Diphtheria Hypothyreosis Leprosy

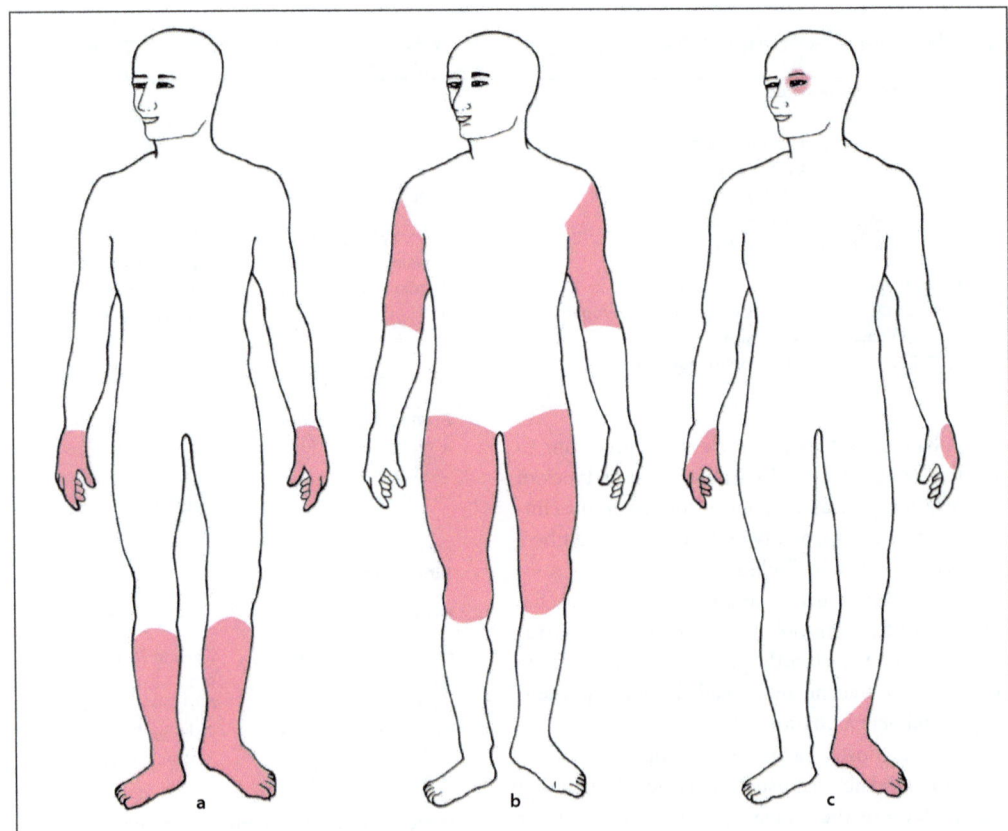

Figure 84 Distribution of motor and/or sensory symptoms in polyneuropathies. **a** Distal stocking and glove symmetrical pattern. **b** The proximal, occasionally asymmetric pattern is rare. **c** The mononeuropathy multiplex affects individual nerves, cranial nerves such as the oculomotor nerve can also be affected

tion phenomena are expected in the examination. For this reason, in electrophysiological examinations, the determination of the NCV plays an important role in diagnosing this form of polyneuropathy.

In demyelinating polyneuropathies, the NCV is affected to varying degrees depending on the type and stage of the disease. The internodal segments are separated by the nodes of Ranvier that notch the myelin sheath wrapped around the axon (Figure 5). Before conduction disorders occur in single-impulse propagation, there can be disruption in the transmission of high-frequency impulse series. The early symptoms of sensory polyneuropathy such as reduction of the sense of vibration or reduction and disruption of esthesia and algesia can be explained as faulty transmission of frequency impulse series.

This information is encoded as frequency impulse sequences by the skin receptors and transmitted to the central nervous system. Insufficient conduction quality leads to an inadequate flow of information. Changing the impulse sequence and changing the single impulse sequences plays an important role in the development of neuropathic pain.

The nerve needs to recover between the transmissions of the single impulses. During this time, it is not excitable at all or shows conduction delays. The time that a nerve needs after stimulation to be able to conduct the next stimulus normally again is called the refractory period and can be measured. In the early stages of demyelinating processes and axonopathies the refractory period may already be lengthened.

However, in an injury of a circumscribed myelin segment, saltatory conduction is disrupted and the impulse transmission can be interrupted locally (Figure 5). As a clinical correlate, there is muscle paralysis, but no EMG denervation phenomena.

Demyelinating inflammatory diseases such as Guillain-Barré polyneuroradiculitis often have a fulminant course. Due to therapeutic interventions such as high-dosage cortisone in conjunction with plasma separation, the prognosis of this acute inflammatory demyelinating neuropathy with respect to restitution is often better than for more chronic axonopathies, which offer few approaches for therapy. After remyelination, the nerve only occasionally regains its original conduction velocity.

The significance of demyelinating processes in pain transmission has already been discussed in the corresponding chapter.

8.1.1 Polyneuritis and polyradiculitis

The inflammatory lesions are important for a differential diagnosis, as their symptoms take many forms and are only rarely mononeuritic, but instead usually polyneuritic. Often the initial event that triggered the inflammation, an infection, for example, occurred so far back in time that the neurological consequences, neuritis as an immune reaction sometime around the time of the initial event, cannot be assessed. There can even be clear differences in the clinical picture, such as for herpes zoster, in which the skin lesions in a classic monoradicular pattern in conjunction with severe pain color the clinical symptoms to such an extent that the sensory and motor deficits are of secondary importance. On the other hand, sensory and motor pareses are the main symptoms in other polyneuritis forms and pain is more of a secondary symptom. In inflammatory nerve affections it is often difficult to distinguish clinically between a radicular polyneuritis and one located further to the periphery. Herpes zoster radiculitis, which actually should be discussed here, is treated in this book along with the pain syndromes because of the domination of the pain symptoms. However, two other clinically and diagnostically significant inflammatory diseases are presented here.

Guillain-Barré polyradiculitis

Polyradiculitis is an inflammatory disease affecting primarily the peripheral nervous system. In addition to humoral factors, cellular immune processes play a role in its pathogenesis. In addition to the acute course (AIDP), whose best-known form is Guillain-Barré syndrome, there are chronic forms, also called chronic inflammatory demyelinating poly(radiculo)neuropathy (CIDP), which have a very heterogeneous etiology.

The Guillain-Barré syndrome can become manifest at any age. It is usually preceded by a bland (gastrointestinal infection) in the patient history 14 days previously; other known triggering factors are: serum transfusions, vaccinations, trauma, operations, and poisoning.

The diagnosis is made from the clinical course of a rapidly progressing sensorimotor polyneuropathy with diffuse pain and the typical cerebrospinal fluid finding, which shows albuminocytological dissociation. This means that the cerebrospinal fluid at the beginning of an inflammatory process has a medium total protein level and in the further course, an elevated total protein level with normal to slightly elevated cell count. Clinically, within hours or a few days, there is rapidly ascending paresis with a loss of reflexes and varying degrees of sensory disorders. Paresthesias can be the first symptom. Involvement of cranial nerves is not rare. Feared complications are paralysis of the respiratory muscles and vegetative function disorders with circulatory dysregulation and cardiac arrhythmia with various degrees of AV blockage. An external demand pacemaker may be necessary to prevent sudden asystoles. About 1–2 weeks after its start, the disease reaches its climax. Depending on the severity of the nerve lesions, restitution ranges from complete remission within weeks to severe defect healing. Acute courses have a better prognosis than chronic ones.

Subacute courses are always difficult to recognize if the patient complains of uncharacteristic symptoms such as fatigue and dysesthesia in the form of myalgia and lumbago. Moreover, focal demyelination with a conduction block can imitate peripheral nerve or root lesions, a special form manifests itself as primary axonal damage.

Prompt plasma separation within the first days or intravenous immunoglobulins 2g/kg over 2–5 days improves the prognosis markedly in acute forms, in particular, it shortens any intensive medical measures.

Borreliosis

Acute borreliosis, which can cause polyneuritis, has recently been observed more frequently. Because it is considered more often as a differential diagnosis of unexplained pain, it is diagnosed more frequently, which results in the apparent increase of frequency of the disease. Borrelias are transmitted to humans by tick bites, in which the tick must have been actively sucking blood for many hours. In Germany an infection rate of about 1% after a tick bite is assumed. An erythema migrans can develop at the bite location. First there is a reddish swelling at the site, which then enlarges to a ring erythema with a diameter of up to 20 cm and then fades over the period of a month from the center to the periphery. In addition to local itching, febrile temperatures, myalgia, and arthralgia can occur. Not all patients can recall such a skin process when pain and paresis develop in the form of mononeuritis after 5–6 weeks, which then usually spread to other nerves and roots. This stage is described as acute neuroborreliosis. Only the joints and/or the heart can be primarily affected. Burning, piercing, and, tearing pain then occurs, usually first in the area around the bite. The pain does not respond well to the usual analgesics. Depending on the extent of the inflammation, there may be more or less pronounced motor and sensory radicular lesions, in which at times severe concomitant meningitic disease justifies the name mengingopolyneuritis (Garin-Bujadoux-Bannwarth syndrome). There are repeated cranial nerve deficits with disorders of oculomotor function and bilateral facial nerve paresis.

The peripheral and central neurological symptoms seem to be triggered by angiitic processes and a local endoneural immune reaction. In North America borreliosis is triggered by the spirochete Borrelia burgdorferi; in Europe also by B. afzelli and B. garinii. It is very interesting that the course of borreliosis is different in North America. The disease is also called Lyme's disease, after the US city where the first epidemic occurred. Immunologically speaking, B. burgdorferi has a somewhat different antigen structure than the European vari-

ants. This peculiarity is presumably the cause of the more pronounced neurological symptoms of patients in Europe; in North America, arthralgia and carditic symptoms predominate.

Unexplained conditions of generalized pain, malaise, myalgia, arthralgia, and fatigue are often found in association with chronic borreliosis. Serological examinations are often misleading and can be interpreted properly only with the results of other examinations. In fact, chronic infection is quite rare and shows, in addition to changes in the cerebrospinal fluid, signs of neurological disorders such as spasticity, myelitis, or encephalitis as well.

The most important diagnostic step is to think of the clinical picture. In the lumbar puncture, there is always an inflammatory cerebrospinal fluid constellation with elevated total protein level and lymphocytic cell increase ranging up to 2,000/3 cells. The IgG and IgM antibody titers must be determined in serum. If a titer is pathologically high, effective antibiotic therapy must be initiated. A 2-week course of penicillin G, 20–30 mega units/day or 200 mg/day of ceftriaxone or cefotaxime is appropriate.

8.2   Axonal degeneration

It is not yet certain whether neuroaxonal disorders can actually be divided into subgroups. But since the concepts associated with subdivision allow useful insights into the pathophysiological processes of these neuropathies, they are presented here separately.

8.2.1   Sensory Neuronopathies

Sensory neuronopathies are a small subgroup of neuropathies. In terms of pathophysiology, there is a primary degeneration of the dorsal root ganglion. Therefore this form of neuropathy is also called sensory ganglionopathy. In neuronopathies, the lesion occurs on the cell body of the nerve. As a result of this lesion, there is secondary degeneration of the dendrites and the axon. It must be considered that the neuron of a sensory nerve is located in the spinal ganglion. From there, one dendrite proceeds to the peripheral system and another centrally to the spinal cord. In cases such as this, there is always a disruption of nerve conduction, a deafferentation (see Figure 7a), toward the central nervous system in addition to the peripheral neuropathy.

Clinically, it comprises a subacute starting progressive symptomatology with an asymmetric, non-length-dependent sensory impairment. Loss of proprioception induces an impaired sense of position, expressed by uncertain gait and balance, ataxia. Owing to the affected neurons of small and medium size, positive sensory symptoms arise such as burning pain, hyperesthesia, and allodynia. Unlike in length-dependent polyneuropathies, the symptoms are not symmetrical but asymmetrical and multifocal. The trigeminal sensory ganglion neurons can also be affected and cause sensory problems in the face.

These neuronopathies are mainly induced by paraneoplastic processes and are often associated with anti-Hu antibodies. A toxic effect can also be seen with platinum-based chemotherapy. Herpes zoster is an example within the causative group of infections. Vitamin B_6 induces a dose-dependent sensory neuropathy/neuronopathy. Vitamin B_6 doses of more than 200 mg/day are critical. A high percentage of patients with Sjögren syndrome have an immune-mediated neuronopathy.

8.2.2 Axonopathies

Axonopathies are the result of direct, primary damage to the axon. The existence of an axonopathy is usually an indication of a poor prognosis. The cause of the axonal neuropathy is generally a disruption of the axoplasmic flow. This axoplasmic flow proceeds distally from the nerve cell and in reverse from the distal end back to the nerve cell. It is substantially involved in the trophic supply of the axon. When there is a disruption of the axoplasmic flow, the first parts of the nerve to be damaged are the ones located furthest distal and these are normally the feet. This is followed by functional failure and dying off of the nerves from distal to proximal, a process known as "dying back."

In Chapter 1 "Basic anatomy" the properties of the neurons are decribed. So neurotransmitters,

proteins, in particular structure proteins, and phospholipids are transported in a fast and a slow fraction in proximo-distal direction within the axon. Disorders in the migration speed of substances can occur as well as qualitative and quantitative changes in the substances present in the axoplasm. The recoupling from the periphery to the center (cell nucleus) is attributed to an additional retrograde flow of axoplasmic transport. Depending on the primary manifestation of the damage, it is called a proximal or distal axonopathy. Most frequent is a distal location, in which the nerve damage progresses slowly from distal to proximal, a process called the dying-back phenomenon (Figure 3). If it is possible to eliminate the damage causing the neuropathy, the nerve must then resprout in order to regain its original function. The sprouting speed depends on how quickly the axoplasmic flow takes place. The slowest component of the flow is estimated to be one millimeter per day. This physiological factor limits the speed of sprouting. In cases in which severe damage to the nerve is present, it must be assumed that this speed is even less. An improvement of neurological deficits cannot occur until the nerve reaches its target organ, for example, the muscle. When there is severe damage, the motor nerve often cannot reach the muscle and the sensory nerve cannot reach the target organs in the skin. Muscles that have been denervated for more than 2 years have most likely been surrounded by connective tissue and are no longer accessible for reinnervation. In an EMG test, lack of pathological spontaneous activity in an axonal lesion combined with polyphasic and long-duration action potentials of motor units are signs of neurogenic regeneration.

Electrophysiologically, disorders in the region of the end branches of motor nerves can be determined by measuring the distal motor latency. Proximal lesions in the root or plexus area are best diagnosed with the H-reflex time or F-wave latency. Occasionally, the sensory evoked potentials can also be useful. These methods are always used when the usual measurement of NCV shows normal values, but there are clinical deficits.

Cytostatics, especially vinca alkaloids, cause axonopathies. The large-caliber axons are especially vulnerable, the small-caliber axons are rarely affected. This selective damage pattern, related to the size of nerve fibers, applies to both motor and sensory nerves. Whether there are actually axonopathies, or instead primary damage in the cell body itself, must be clarified by further research. The fully regenerated nerve never achieves the original conduction velocity. As a morphological cause of the reduced NCV, smaller axon diameters or shorter internodes are found on the nerve.

8.3   Combined axon and myelin damage

For the theoretical discussion of the development of a polyneuropathy, the differentiation into purely axonal and purely myelin damage is important, but in reality it is rarely that clear-cut. Axonal lesions often very rapidly affect the myelin as well through the impairment of trophic supply to the axon. In genetic polyneuropathies, for example, in Charcot-Marie-Tooth disease, a neuronal form or a hypertrophic form, marked by hypertrophic, onion skin-like layers of myelin formations with low to absent NCV, can be formed. Severe demyelinization from inflammatory infiltrates such as in Guillain-Barré syndrome can also damage the axon membrane and thus trigger denervation phenomena in the muscle. The course of diabetic neuropathies always includes a mixed form of neuron and myelin damage. In addition, nerve infarctions in the form of an ischemic lesion can occur as a complication of diabetic microangiopathy. As a final result of this disease, there is almost always a combined lesion. In view of the variety of noxa that can lead to polyneuropathy, thorough clarification of the etiology is necessary. Despite an intense search, the etiology of over 30% of polyneuropathies remains unclear. Targeted intervention is possible only if the etiology is known. It has been shown again and again that avoidance of the damaging noxa, such as alcohol abuse or an elevated blood sugar level, is the most effective therapy.

8.4   Diabetic neuropathy

Diabetes comprises a group of metabolic diseases. Serious long-term complications, besides those of the peripheral nervous system, include stroke, foot ulcers, amputation, cardiovascular disease, chronic kidney failure, and damage to the eyes. Diabetes can afflict single nerves as a cranial or peripheral mononeuropathy or mononeuritis multiplex, but it generally causes a sensorimotor and autonomic neuropathy. The symptoms usually develop in a length-dependent manner, starting distally and extending proximally. Numbness, pricking, itching and tingling, pins and needles sensations, and/or burning pain starting in the feet are the first presenting symptoms. Walking is often described as "walking barefoot on hot sand." Patients also sometimes complain of muscle cramps. As the symptoms can worsen during night, the condition interferes with sleep and affects enjoyment of life and daily activities.

This polyneuropathy is complex, as diabetes can affect large myelinated and small nerve fibers to different degrees in different individuals. Characteristic morphological changes are:
- "Dying back" phenomenon with loss of axons (Figure 3)
- Demyelination in focal areas
- Reduction in myelinated fiber density

A single clinical examination or electrophysiological test cannot diagnose all these changes. Nerve conduction tests can only disclose large fiber damage. Loss of ankle jerk and diminished vibration sense are also not always present. Small fibers (thinly myelinated A-δ-fibers, unmyelinated C-fibers) constitute more than 80% of the peripheral nerve. The small fibers have autonomic functions such as the regulation of sweating or blood pressure and additionally conduct sensations, such as temperature, itch, touch, and pain. Autonomic impairment leads to dry mucosa, or changes in micturition. Owing to neuropathy of the parasympathetic and sympathetic nervous system innervation of the corpora cavernosa and of the dorsal penis nerve, diabetic men develop erectile dysfunction.

Malfunction of nerve fibers induces positive and negative signs. Tests have to be used to assess the status of broad symptoms. A small fiber neuropathy can be assumed when functional of morphological impairment is detected in the following tests:
- Neurological examination
- Quantitative sensory testing (QST)

▬ Morphometric quantification of intraepidermal nerve fibers (IENF)

To diagnose the presence and etiology of a small fiber neuropathy, a pragmatic work-up is needed so as to optimize treatment, which is mainly symptomatic, as outlined in a review by Chan et al. (2016, *Muscle & Nerve* 53: 671).

Questionnaires should be used as a first step to obtain a detailed pain history. Punch skin biopsies are minimally invasive procedures and allow for IENF. Biopsy specimens should be obtained from standardized sites like the lateral lower leg, for which published normative values are available. A noninvasive method is corneal confocal microscopy, with which visualization of the sub-basal nerve plexus in the cornea is possible. Published data show a good correlation of this technique with IENF results and disease stage.

The number of diabetic patients who have polyneuropathy varies substantially between 29 and –75%. Polyneuropathy manifests earlier in type 2 than in type 1 diabetes. One third of these patients suffer from chronic painful neuropathy. The results depend on the accuracy and implemented examination methods. The most important risk factor for developing neuropathy seems to be poor glycemic control. However, it is unclear whether there is a glycemic threshold beyond which nerve dysfunction develops.

If there is rapid change in glycemic control as a sudden improvement or poor control, an acute painful peripheral neuropathy can evolve. Typically it is seen with fast and successful treatment in acute decompensation. In regaining glycemic control this acute form has a good prognosis to fully disappear within a year.

The importance of hyperglycemia for the development of diabetic polyneuropathy has been proven. In population-based studies, 25–60% patients with diagnosed idiopathic neuropathy were reported to have prediabetes. There is a lifetime risk of about 50% that patient with prediabetes develop type 2 diabetes. In people with prediabetes, up to 25% had a neuropathy and among them about 20% had neuropathic pain. The mechanisms by which neuropathy comes into being is still unclear. Hyperglycemia seems to induce the mitochondrial production of free radicals. This induces oxidative and nitrosative stress. A consistent feature common to all cell types is that they are damaged by hyperglycemia. The increased production of reactive oxygen species (ROS) plays an important role. When increased intracellular glucose generates increased ROS in the mitochondria, these free radicals can induce DNA strand breaks and alter cell metabolism. Moreover, hyperglycemia activates the polyol pathway and increases sorbitol concentration in nerve cells. This changes the osmotic gradient, leads to Schwann cell damage and to myo-inositol deficiency, and affects the electrophysiological stability of the nerve cell membrane. In addition, microvascular pathology can change the nerve's blood supply and cause endoneurial hypoxia and ischemic nerve damage. Unfortunately, we do not have biomarkers by which we can predict the risk of developing neuropathy or neuropathic pain. Despite the known link between hyperglycemia and development of polyneuropathy, it is not proven that low glucose protects against the development of polyneuropathy. For type 1 diabetes, glucose control has proven to be effective, in type 2 diabetes, however, it unfortunately has not. It certainly is important to achieve a good glucose control in diabetic patients in order to influence the multiple comorbidities – for example, nephropathy, retinopathy, peripheral artery and cardiovascular disease, and depression – associated with this metabolic disease. But studies have not proven that these interventions have a favorable effect on the development and progression of diabetic polyneuropathy. There are also theories that hyperlipidemia correlates with myelinated fiber loss, independent of disease duration, age, or diabetes control.

Pain in diabetes may arise not only from changes in the periphery. Vascular and cellular changes also take place in the central nervous system. This hints to the possibility of central sensitization as an additional pain source and would be one reason for explaining the problems in treating diabetic chronic pain.

Treating diabetic polyneuropathy has two aims: to achieve a better control of the metabolic situation and to alleviate painful symptoms.

The treatment of pain in diabetic polyneuropathy follows the proposals as given in Sect. 10.3.1.

For diabetics, potential weight gain can be a concern with tricyclic antidepressants. This medication with anticholinergic side effect may also induce orthostatic hypotension, especially in those patients who suffer from autonomic neuropathy. In diabetic patients, the parasympathetic fibers are affected earlier than the sympathetic fibers. Therefore the reflex responses to changes in hypotension are not compensated by adequate cardiac output. Myocardial infarction in diabetic patients is usually more extensive and more severe than in nondiabetic patients. Asymptomatic myocardial infarction or asymptomatic myocardial ischemia occurs more frequently in diabetic patients. Reasons for this are different thresholds of pain sensitivity, psychological denial, or the presence of autonomic neuropathy leading to sensory denervation of the heart muscle. The latter seems to be more likely in diabetic patients, because autonomic neuropathy is a common feature of diabetes, and abnormalities of the autonomic nerve fibers were demonstrated histologically in diabetic patients who died after painless myocardial infarction.

Diagnose and treatment is needed in diabetes as early as possible. This is the best attempt to prevent multi-organ damage. If profound morphological damage has happened to the peripheral nerve, a so called "point of no return" is reached. This is the case when about more than 50% of fibers within a nerve are damaged (see Fig. 83, D). Then the likelihood for regeneration in a nerve is low. In this case clinical evidence shows no chance for improving the lost nerve function.

Pain has a major negative impact on the quality of life. One reason is that the patients develop depression and anxiety. Time and again, pharmacotherapy of painful diabetic neuropathy has proven to be ineffective. The additional use of antidepressive medication along with psychosocial assessment and evaluation of coping skills may be of great utility for unresponsive patients. Catastrophizing is a very strong psychological predictor for worsening of pain experience. When patients expect to feel worse, they eventually tend to do so. To learn and use coping techniques can be helpful for these patients.

Clinical aspects of painful conditions

9.1   Complex regional pain syndrome

Sudeck's atrophy, also known by the synonym "reflex sympathetic dystrophy," is now described by the term "complex regional pain syndrome" (CRPS). This umbrella term has been subdivided into type I and type II. CRPS I is intended to cover reflex sympathetic dystrophy and similar disorders without a nerve injury, while CRPS II occurs after damage to a peripheral nerve or plexus and was formerly called "causalgia."

CRPS is not an entity but a diagnostic term to describe a great variety of disorders. The syndrome is diagnosed purely on the basis of clinical signs and symptoms. All forms have in common a spontaneous or stimulus-induced pain. Compared with the normal healing course of a similar trauma, the pain in CRPS is disproportional in magnitude and duration to the inciting event. The clinical pictures are shaped by a myriad of autonomic, sensory, and motor disturbances in highly variable combinations. Owing to the diversity of symptoms and variable pathophysiology, different diagnostic criteria were published. To have a modern and more standardized approach to the diagnosis of CRPS, the journal Pain published the "Budapest Criteria" (Table 14) in 2010. They are totally dependent on the subjective response of the patient and the subjective impression of the clinician and no further diagnostic procedures are needed. There are currently no tools for diagnosing CRPS but there are some helpful tools for confirming the syndrome.

CRPS develops as a localized pain disorder about 4–6 weeks following a trauma. It manifests in about 7% of patients who have limb fractures, limb surgery, or other injuries to an extremity. The disease peaks in people aged between 50 and 70 years, and women are three to four times more affected than men. Primarily the upper limbs are affected. Initially after a fracture, limbs can exhibit CRPS but symptoms often resolve spontaneously. The time window for the onset of CRPS after the initial injury is usually 3–4 months. A diagnosis of CRPS after many months or years following a trauma should be questioned.

The spontaneous and evoked regional pain character is described as pricking, burning, and attacks of shooting painful sensations localized deep in the tissue and especially at rest. As pain immobilizes the limb, disuse effects will quickly jeopardize function and mobility. The pain arises from sensitization of deep somatic primary afferents. Allodynia indicates central sensitization. Blunt pressure on muscles or bones is felt as painful. Cold skin can be

Table 14 Budapest clinical diagnostic criteria for CRPS, codified by the International Association for the Study of Pain[a]

1 Continuing pain, which is disproportionate to any inciting event.

2 Must report at least one symptom in three of the four following categories:
   - Sensory: reports of hyperesthesia and/or allodynia
   - Vasomotor: reports of temperature asymmetry and/or skin color changes and/or skin color asymmetry
   - Sudomotor/edema: reports of edema and/or sweating changes and/or sweating symmetry
   - Motor/trophic: reports of decreased range of motion and/or motor dysfunction (weakness, tremor, dystonia) and/or trophic changes (hair, nail, skin).

3 Must display at least one sign at time of evaluation in two or more of the following categories:
   - Sensory: evidence of hyperalgesia (to pinprick) and/or allodynia (to light touch and/or deep somatic pressure and/or joint movement)
   - Vasomotor: evidence of temperature asymmetry and/or skin color changes and/or asymmetry
   - Sudomotor/edema: evidence of edema and/or sweating changes and/or sweating asymmetry
   - Motor/trophic: evidence of decreased range of motion and/or motor dysfunction (weakness, tremor, dystonia) and/or trophic changes (hair, nail, skin).

4 There is no other diagnosis that better explains the signs and symptoms.

a All four categories have to be fulfilled for the diagnosis of CRPS for scientific study purposes. For clinical purposes, at least two of the clinical signs in Category 3 have to be present.

CRPS, complex regional pain syndrome.

a sign of a sympathetic involvement. In the acute stage there is limb edema with red and hot skin. This can change into cold limb temperatures with livid blue skin and increased sweating. The patterns of vasoconstrictor hypoactivity and sudomotor hyperactivity may be a sign of sympathetic dysfunction or possible sympatho-afferent coupling (Figure 82). Edema can also be caused by inflammatory mediators such as bradykinin and pronociceptive neuropeptides such as substance P and cytokines. These substances affect blood vessels and can induce an increase in plasma extravasation and vasodilatation. The clinical signs are sweating and red skin color and "hot" temperature asymmetry in the acute phase. These mechanisms result in a typical temperature difference between the contralateral and affected limb of more than 1°C. This initial phase of the disease is often called "warm complex regional pain syndrome." The cyclooxygenase pathway can enzymatically induce inflammation, but an oxidative stress pathway can also induce it nonenzymatically. All this is not static. Patients with either warm or cold skin react differently if they are exposed to cold. Adaptation to warm ambient temperatures equalizes the aforementioned temperature difference in the limbs. Thus temperature and stress seem to induce centrally regulated reflex mechanisms, which overreact with warmth and underreact with cold in the skin vasoconstrictors. Owing to the autonomic changes, hair and nail growth are reduced and contractures may develop within connective tissue. Additionally, motor changes can be observed from a weakness of the affected side with brisk reflexes up to tremor or dystonia.

The diagnostic yield of the implemented tools depends on the stage of the disease, if it is acute or chronic. It also depends on whether it is an inflammatory peripheral or a central CRPS type with or without additional neuropathic or sympathetic pain features. Plain X-rays show patchy osteoporosis, but this is insensitive. X-ray imaging is only useful if both hands are on one radiograph. Three-phase bone scintigraphy can be a useful tool for confirming CRPS, allowing for a quantitative analysis of the region of interest. The finding is positive with an increased radiotracer uptake in the mineralization phase in the joints as opposed to the joints that are not affected by the initial trauma. However, a nega-

tive finding does not exclude CRPS. Quantitative sensory testing is not suitable for making a diagnosis, but pressure pain over the bones and joints supports CRPS.

It still is a puzzle what the important etiological pathophysiological events are that alone, in sequence, or in interplay are responsible for the development of the diverse representations of CRPS. Under research and discussion are the following multifactorial process involving both peripheral and central mechanisms: Inflammation is a hallmark and persists for months, abnormal cytokine production sensitizes the peripheral and spinal nociceptive system, and autoimmune responses generate autoantibodies to adreno- and acetylcholine receptors and provide a link to the sympathetic nervous system. Altered blood flow in skin vessels leads to cold skin. Skin temperature was therefore taken as a classifier to differentiate cases directly after the trauma into primarily cold and warm cases. In the chronic stage, an inhibited sympathetic nervous system could sensitize adrenoceptors in blood vessels, lead to vasoconstriction, and contribute to cold extremities. Inflammation-induced overstimulation of fibroblasts leads to articular contractions. Ongoing noxious input increases the excitability of nociceptive neurons in the spinal cord and leads to central sensitization. In later stages, motor symptoms can become prominent as fixed dystonia and reduced limb positioning accuracy. This has been shown to be disturbed sensory-motor integration with central cortical reorganization and distorted subjective mental representation of the affected limb. Brain imaging has shown reduced limb representation in both primary and secondary somatosensory cortices in patients with upper limb CRSP. In the insular and cingulate areas responsible for the affective components of pain, reduced gray matter volumes have been detected. All this causes impairments in the patient's daily living and ability to function.

The therapy for CRPS should reflect the pathophysiological findings. There are no positive clinical studies proving that medication – including opioids, antidepressants, and anticonvulsants – is helpful, which is usually used in neuropathic pain. In the acute stage of CRPS with prevalent tissue edema and inflammation, cortisone has a positive effect. Prednisolone at 100 mg for 4 days can be given, with the dose tapered by 25 mg every day. Cortisone can prevent neurogenic inflammation and neuropathic pain. Hypofusion of tissue and generation of free oxygen radicals due to hypoxia are the rationale for the use of radical scavengers such as 50% dimethyl sulfoxide (DMSO) cream, applied about four times per day. A further therapeutic trial should be performed with calcitonin and diphosphonates for 1–2 weeks to affect bone metabolism. For pain control, continuous sub-anesthetically dosed IV infusion of ketamine has been reported to be helpful but serious side effects can also occur. The role of 500 mg vitamin C daily as an antioxidant to prevent or reduce inflammation due to oxidative stress is unclear.

In the past, the involvement of the sympathetic nerve was believed to be highly important and the main etiologic factor for CRPS. Therefore this pain was called sympathetically maintained pain (SMP). This is in contrast to other forms of pain described as sympathetically independent pain (SIP). A neither – nor differentiation is not possible. Different diseases such as phantom limb pain, zoster neuralgia, and CRPS can consist partly of SIP and SMP. Proof for the sympathetic involvement in CRPS was thought to be a positive response to stellate ganglion blocks or a lumbar sympathetic block. But a Cochrane Database Systematic Review in 2013 and again in 2016 (Cochrane Database Syst Rev. 2016;28(7)) disclosed a scarcity of published evidence supporting the use of local anesthetic sympathetic blockade for CRPS to reduce pain. Therefore, these blockades should no longer be used, except for a small subgroup of patients with definite SMP.

In CRPS there also seems to be incongruence between motor output and sensory input. A new treatment option uses "mirror therapy." Cognitive and behavioral treatment strategies are used to restore distorted brain circuits. Thereby, congruent visual feedback from the moving unaffected limb is used to exercise the affected limb. This visual input from a moving unaffected limb can reestablish a pain-free relationship between sensory feedback and motor execution in the affected limb. Physical therapy is therefore essential but must follow the rule of not causing too much pain. A combination of passive maneuvers with active movements must be guided by the patient's pain threshold.

Therapy in the past advocated rest and inactivity for the affected limb. The worry of causing damage through movement dominated physiotherapy concepts. Today new concepts have arisen. Pain-exposure physical therapy is a progressive-loading physical exercise program and management of pain-avoidance behavior. Another program is called graded exposure in vivo. The aim is first to reduce irrational disease-related fears and the next step is behaviorally oriented identification of fear-related tasks. In coping with these tasks the patients is expected to become less afraid of performing these activities. If a relevant mental health comorbidity is present, the patient should be referred for psychotherapeutic treatment.

9.2   Root avulsions

After root avulsions or arm plexus lesions, extremely unpleasant pain occasionally occurs in denervated and anesthetic dermatomes that is refractory to conservative treatment. The site where pain originates is seen in the spinal cord, and the loss of any sensory afferents from the periphery (deafferentation) plays the most significant role. Let us recapitulate that the central axons of the spinal ganglia proceed via the dorsal root to the spinal cord and take up synaptic contact with central pain transmission neurons in the dorsal root entry zone (DREZ). Root lesions can destroy only the peripheral segment of spinal ganglion cells. The spinal ganglion can become hyperactive due to the absence of afference and develop spontaneous ectopic activity, thus sensitizing neurons in the dorsal horn. If the entire afferent neuron is destroyed, afference from the periphery is also lacking, which exerts an inhibiting influence on the 2nd spinal nerve via interneurons, thus promoting its hyperactive discharge and sensitization, which is then conducted to the central system. The DREZ operation (Figure 85) attempts to destroy the afferents from spinal ganglia and neurons in the dorsal horn with thermal or laser coagulation, a procedure that often brings about a clear improvement of deafferentation pain.

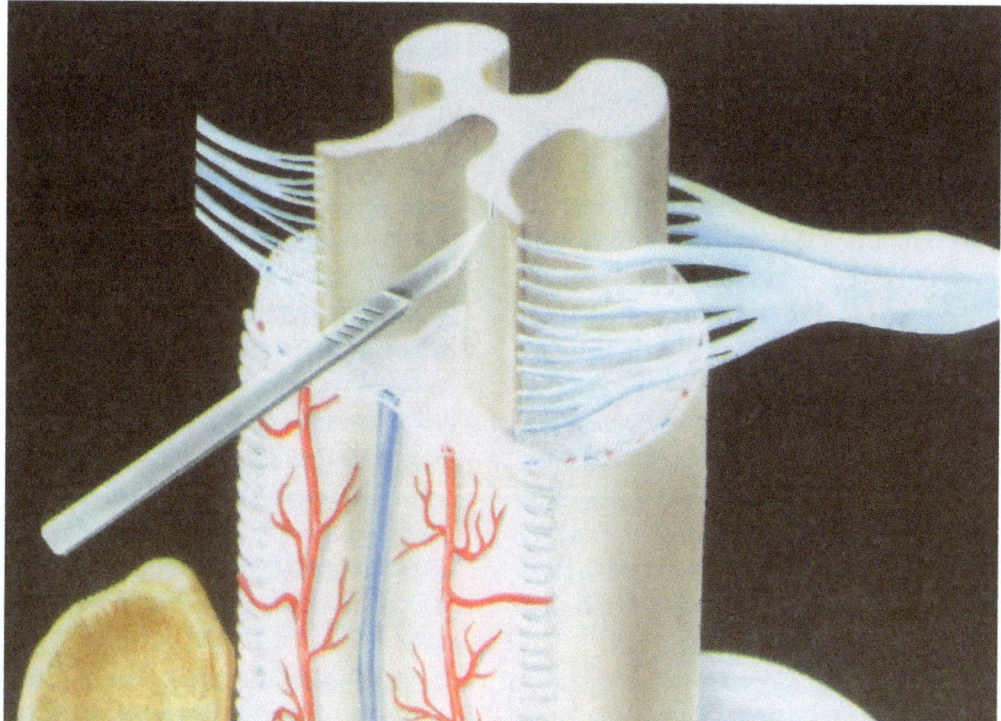

Figure 85 Scheme of a dorsal root entry zone operation

9.3 Referred pain

Referred pain is an experimentally proven physiological phenomenon in humans. This pain is very significant because it is based on mechanisms that help explain a variety of pain states. Although they do not correspond to the actual situation in every case, conceptual, diagnostic, and therapeutic models have been derived from this concept that now occupy a useful, firm place in pain therapy. In the same manner, the gate control theory described above made a fruitful contribution to understanding pain mechanisms and brought pain mechanisms to center stage, although the physiological postulates it is based on did not prove to be sustainable.

For understanding referred pain, it must be remembered that pain can arise in the skin, visceral, or musculoskeletal system. The processes in the skin are best examined, because they are the most easily accessible. Knowledge of the other systems is incom-

plete. Clinically, various pain types are known according to the different sites of origin of pain. While pain from the skin is rapid, sharp, and can be easily localized, musculoskeletal or visceral pain is dull, with a slow beginning and imprecise localization. Pain localization is also conspicuous in that pain can be perceived in regions at a distance from the site of stimulation. Pain projection from organs to certain areas of the skin has been the subject of many studies and was also examined by Henry Head. He postulated in 1893 that the innervation of an organ and its cutaneous projection area proceeds constantly over the same spinal segment and he coined the term "referred pain." Projection areas and the connection of some organs are shown in Figure 86.

This correlation can also be seen in pain projection from muscular and joint structures, although the constant allocation as in Head's zones does not exist.

In theories on the phenomenon of the Head's zones and the occasional segmental projected pain, a

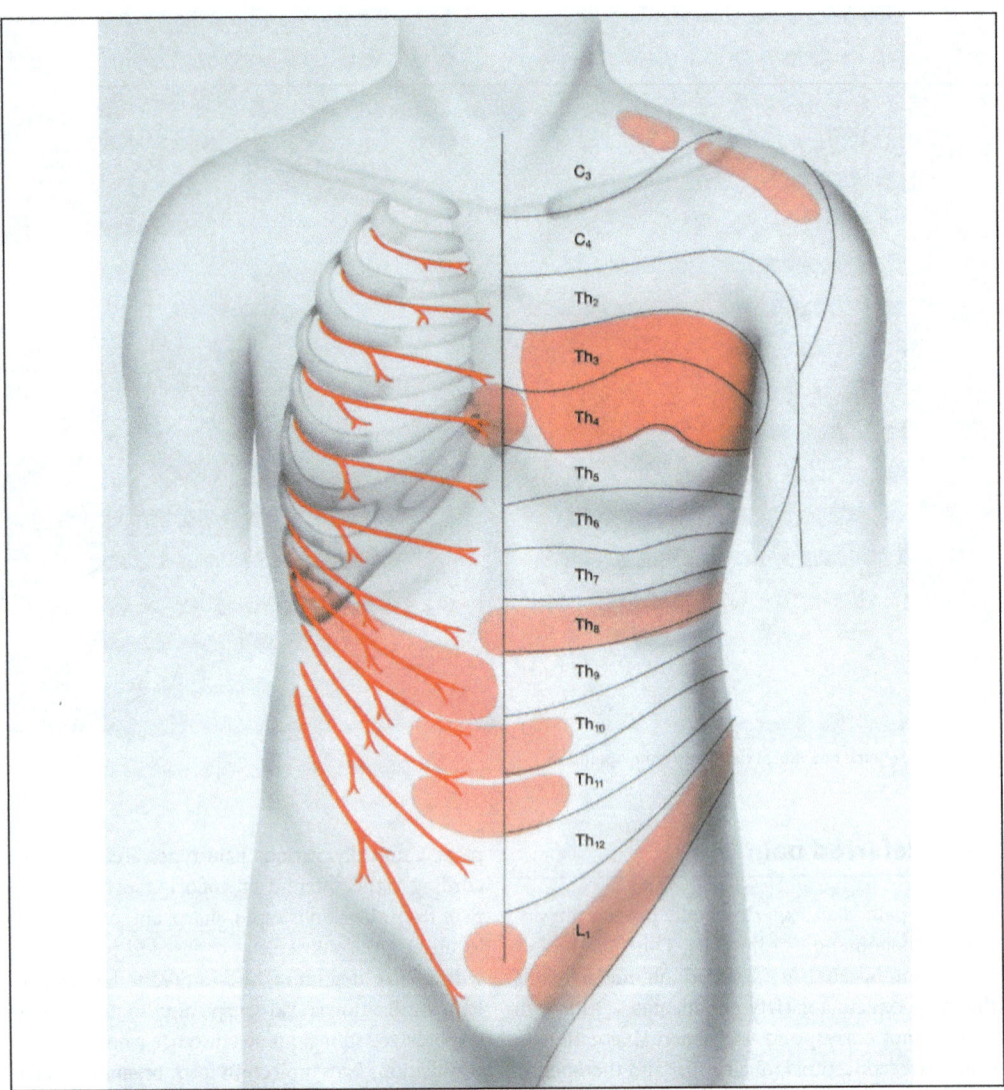

Figure 86 Representation of visceral pain on the skin

link was seen between single afferent transmission paths and their convergence at the spinal level. According to this convergence theory, information proceeds from the organs via viscerosensory fibers allocated to the autonomic nervous system that have a different segmental structure than the sensory system. The afferent impulses proceed via the sympathetic trunk to the dorsal horn, where they connect synaptically to sensory conduction pathways and pain neurons in close proximity to other nerve bundles. Information from two different systems is transmitted over the same pathway to the central system. In the central nervous system, the origin of the information cannot be distinguished. As a result, pain in internal organs is misprojected to areas in which nociceptors from the same segment are located.

Another explanation is that some neurons receive only inputs from the skin or from deep tissue such as muscle and joints. However, many neurons exhibit convergent inputs from skin and deep tissue, and all neurons that receive inputs from the viscera also receive inputs from skin (and deep tissue). Owing to this uncertainty about where the information comes from, pain arising from visceral disease is felt as occurring in a cutaneous or subcutaneous area. Another encoding problem is that, in particular, WDR neurons often have large receptive fields, and a stimulus of a defined intensity may elicit different intensities of responses when applied to different sites of the receptive field.

Processes in the diaphragm (innervated by C2/C3) can thus radiate into the sensory C2/C3 segments at the neck and throat. Another attempt to explain this assumes the existence of dichotomized sensory axons. This involves the splitting of the peripheral axon into several or at least two parts near the cell body. With a peripheral process consisting of several parts, the sensory neuron can than reach the peripheral nerves and musculature via the plexus on the one hand, and on the other hand, reach the skin the same way, but via other nerves. An irritation of the median nerve in the carpal tunnel can thus cause pain and tension in the musculature of the shoulder-neck region. Referred pain can also become manifest as a vegetative reaction in its projection zone. Changes in circulation, either local or in an extremity, lead to temperature fluctuations or piloerection (goose bumps) on the skin.

9.4 Myofascial syndrome

The myofascial syndrome, which, however, is based more on empirical clinical evidence than proven laboratory results, can be considered one form of referred pain. Among the verified knowledge are thus a number of syndromes, which, in addition to individual clinical features, have many things in common so that it is not possible to distinguish sharply between them. This semantic Babylonian confusion includes concepts such as myofascial pain, fibrositis, repetitive motion musculoskeletal problems, repetitive strain injury or work-related chronic myalgia, pseudoradicular syndrome, and low back pain. Fibromyalgia also falls into this category and requires a confirmed and accepted etiology.

These diagnoses just mentioned include a group of pain patients that complain of constant dull,

piercing pain located far from the joints in the musculature. The intensity of pain can change during the day or in the course of the symptoms, often lasting months. The patients have usually consulted several physicians, but all laboratory and technical examinations, including EMG, are unremarkable. Despite its local character, as a differential diagnosis, the pain must occasionally be distinguished from a radicular syndrome, which is why it is sometimes called pseudoradicular syndrome. Although there is no nerve lesion, the pain can cause muscle inactivity and lead to muscle atrophy, simulating a neurogenic lesion.

The diagnosis of a myofascial syndrome often fails due to the lack of understanding of this syndrome and the failure to look for trigger points, which are the leading symptom for making a diagnosis. Trigger points are sensitive soft regions in the muscle that are located in palpable nodules or bands of hard muscle tissue. The trigger points can also be located outside of the muscle in its tendons or their insertion area. Common to all points is that they are very painful when compressed. The pain is initially local, but pain can also be triggered simultaneously in another seemingly unrelated region. The myofascial syndrome is primarily always a local disorder, which, however, can also be seen in systemic diseases such as rheumatoid arthritis or fibromyalgia.

The regional character including certain muscles, also more distant ones, is explained in chirotherapy and osteopathy using the term myotactic unit. It is composed of muscles that function synergistically as agonists or antagonists in a sequence of motions. The significance of the myotactic unit is close to the term tendomyotic chain, often used in chirotherapy. When a muscle develops a trigger point, other muscles can generate other points secondarily, until the entire unit is involved. Table 15 shows indicators for finding trigger points.

This subject has been handled in particular by J. G. Travell and D. G. Simons and covered extensively in their book *Myofascial Pain and Dysfunction*. Using diagnostic local anesthesia of the trigger point in the muscle, pain can often be stopped, confirming the diagnosis of a trigger point. Nociceptive afferents from an area of the body can lead to a reflex-induced, circumscribed muscle contraction

Table 15 Indicators for locating trigger points
1 Soft zones within a myogelosis.
2 Pressure on this area leads to a short, local muscle contraction.
3 Referred pain and vegetative phenomena can be triggered from this area.
4 Limited functionality and weakness of the affected muscle.

(see Figure 89). Other pain-triggering and -sustaining mechanisms are shown in Figures 87 and 88.

The primary myofascial syndrome can occur independently when irritations in the trigger point itself trigger and sustain the muscle contractions via spinal reflex arcs after the elimination of the original nociceptive afferents. However, local muscle contraction from pressure on the trigger point is not a nerve-controlled reflex, but is probably the result of increased mechanosensitivity in the endoplasmic reticulum, the calcium-storing system of the muscle cell.

The functional relationship between joints and the corresponding muscles and tendons can be seriously impaired by mechanical factors such as different leg lengths, occlusion disorders of the teeth, strains in the cervical spine, scoliosis, work-related posture anomalies, or athletic overuse. These are probably the most frequent causes of a primary myofascial syndrome. Systemic disorders can also act as triggers or promote the development of this syndrome. They include in particular thyroid disorders of rheumatic diseases, malabsorption syndromes, chronic infections, and special metabolic situations such as during pregnancy or old age. In this context, the issue of the necessary vitamin supply of the body is often raised. Protective properties for maintaining the function of the neuron are attributed to these substances. They play a role in neuropathies and musculoskeletal pain.

The therapy involves manually stretching the myogeloses, after cooling with chloroethane spray, against the direction of muscle tension, followed by local heat application. Local ischemia in the trigger zone from approximately 30 s of firm pressure on the trigger area is also effective. In addition, stretch-ing exercises of the muscle alternating with isometric contraction can relieve myogeloses.

9.5   Pain states in the trunk and extremities

There is often confusion on how to communicate pain radiating in the back, arms or legs. Pain that extends distally along the arms or legs is clinically divided into radicular and pseudoradicular pain. Radicular pain is defined as a pain state that extends across and below the knee, whereas pseudoradicular pain does not extend further into the leg than the knee. As explained in Sect. 9.3, there is the hypothesis that pain from local proximal disorders that does not affect any nerves or nerve roots, such as facet joint pain or from muscle problems, is felt in a more proximal dermatome as referred pain or in the Head's zones. But pain from disorders associated with nerve root compression is perceived radiating into the distal parts of the dermatomes below the knee. This distinction of radiating radicular pain and pseudoradicular pain is of clinical relevance since a radicular pain always involves damage or irritation of peripheral nerves or nerve roots. This sort of pain is of neuropathic nature, whereas referred pain occurs without nerve involvement and might therefore be nociceptive pain. This distinction is relevant for therapeutic purposes, since a mixture of neuropathic and nociceptive pain, the mixed pain concept (see Sect. 10.3.8), requires other therapeutic strategies than neuropathic or nociceptive pain alone.

Referred pain in myofascial syndromes is elicited from trigger points within muscles or areas around the joint capsule, the tendon insertion, or the periosteum. These painful conditions are also occasionally called tendomyoses. Common trigger points, their anatomical structures, and projection zones are shown in Figures 87, 88, and 89.

In chronic back pain, degenerative processes affect the disc, the bony parts of the spine, and the neuroforamen. Nerve roots may be injured by mechanical pressure (mechanical-neuropathic nerve root pain). Inflammatory mediators from a degenerative intervertebral disc may cause chemical damage to nerve roots in the absence of actual mechanical compression (inflammatory-neuropathic nerve

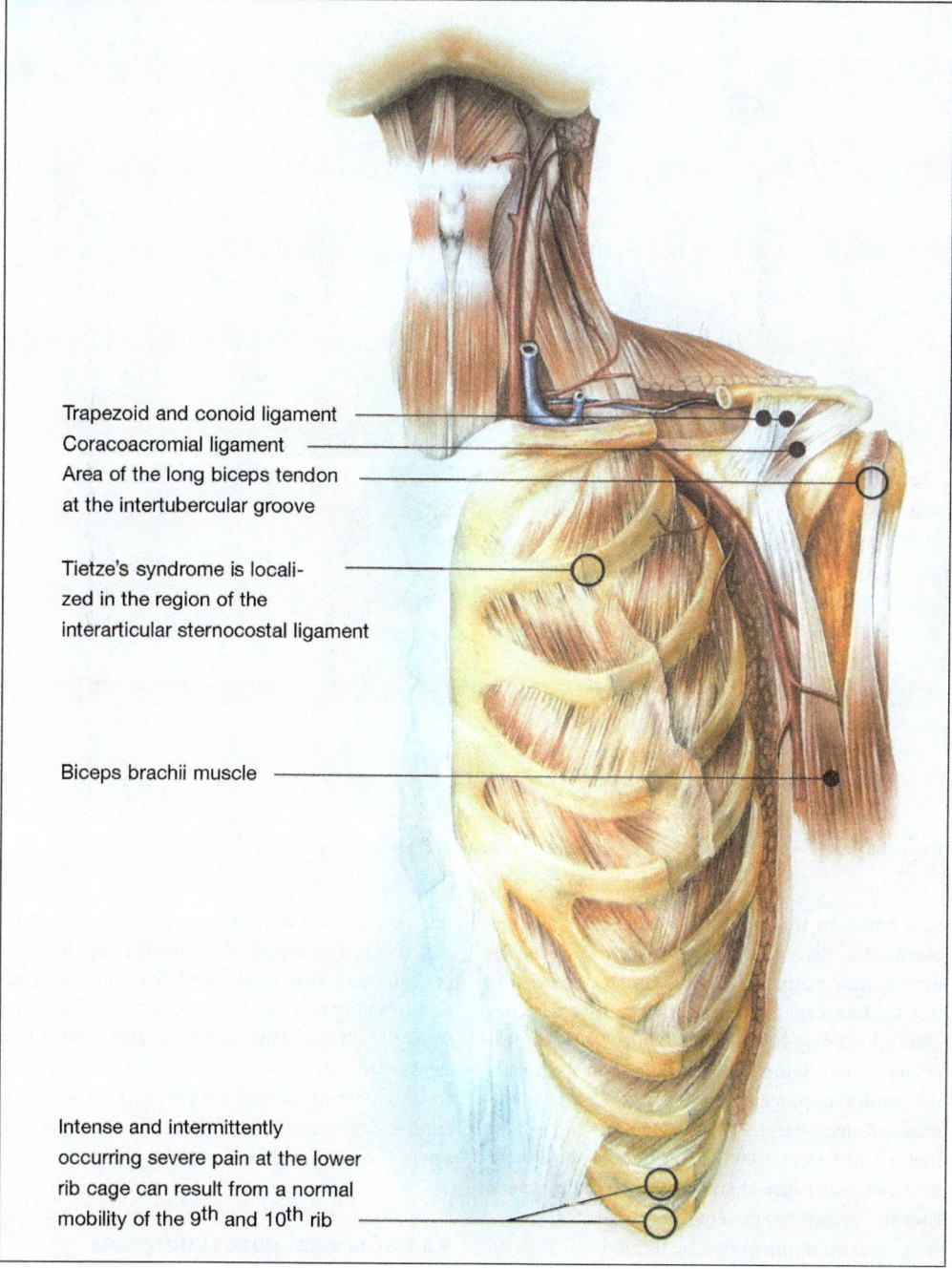

Trapezoid and conoid ligament

Coracoacromial ligament

Area of the long biceps tendon
at the intertubercular groove

Tietze's syndrome is localized in the region of the
interarticular sternocostal ligament

Biceps brachii muscle

Intense and intermittently
occurring severe pain at the lower
rib cage can result from a normal
mobility of the 9th and 10th rib

Figure 87 Trigger point with pseudoradicular pain projection. Trigger zones near joints often elicit pain in muscles with a functional relationship to the joint. This can make it appear that there is a segmental connection

9

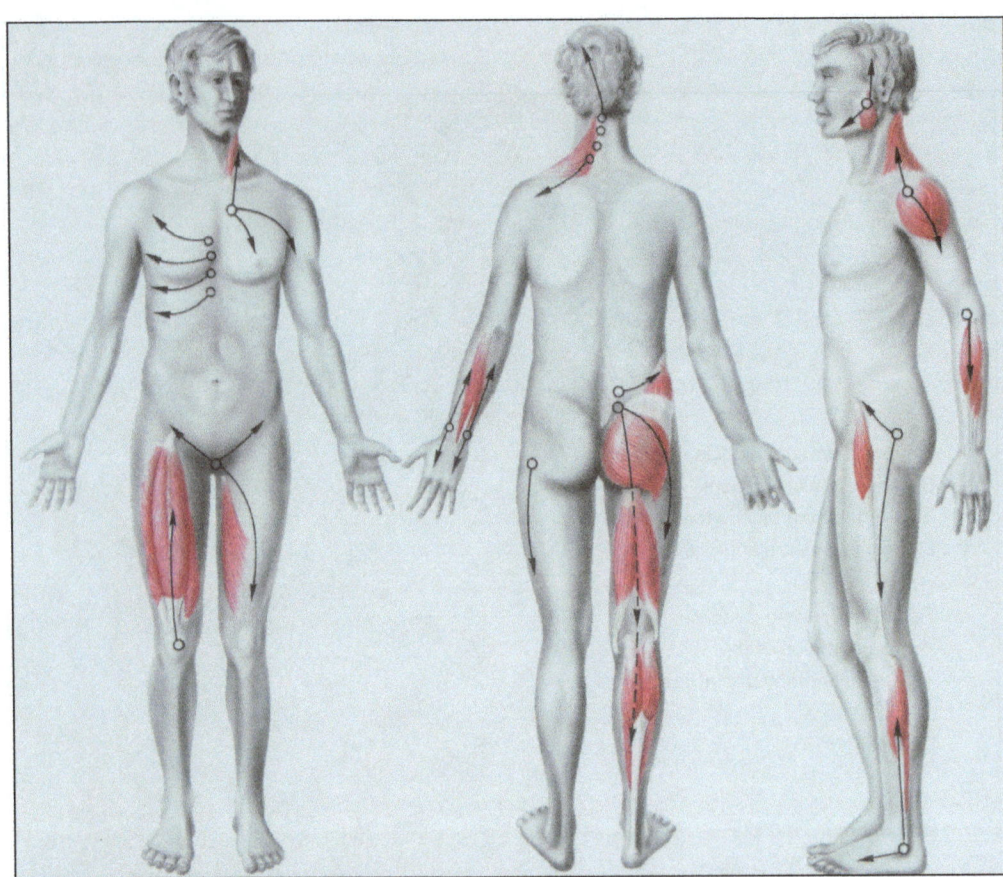

Figure 88 Tendomyopathies and myalgia can arise from irritations in the depicted structures

root pain). In this situation a mixture of noxious events takes place that can generate nociceptive and neuropathic pain. It is important to realize that nociceptors can resprout into the intervertebral disc, which may be damaged by degenerative processes. In addition, the nerve root may be injured by mechanical impingement – resulting in typical mechanical-neuropathic nerve root pain – and inflammatory mediators from the degenerative intervertebral disc may cause chemical damage to the nerve root in the absence of actual mechanical compression. This component can be termed inflammatory-neuropathic nerve root pain. The relative contribution of the various components may differ greatly between individuals, with seamless overlap in many cases. As imaging procedures are not helpful in differentiating between the different pain states,

it is very important to make a detailed history and neurological examination for identifying the mixed pain states. A structural correlate cannot be identified in many cases in patients with all the clinical signs of sciatica. This applies in particular to patients with chronic sciatica. Furthermore, the nerve root may be damaged by inflammatory mediators from a degenerative intervertebral disc in the absence of direct mechanical pressure on the root.

9.5.1 Cervical pain syndromes

Acute painful limitations of the neck musculature are usually a consequence of movement and stress-related muscle trauma with reflectory myogelosis. In contrast to other segments of the

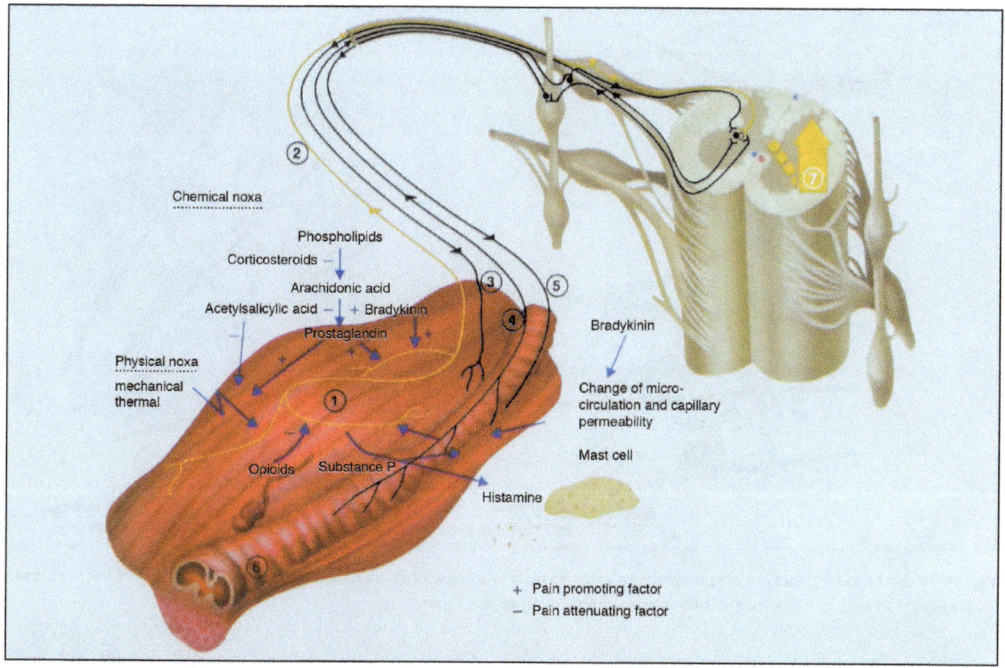

Figure 89 Scheme of mechanisms producing nociceptive pain in muscle tissue. The drawing depicts a muscle with local spasm, a vessel, the spinal cord, and the afferent and efferent nerve fibers innervating these tissues. The nociceptors of the afferent (yellow) nerve convey their message to the spinal cord. There the information is sent by the contralateral spino-tha-lamic tract to the brain. At the same time, reflex mechanisms are activated which control the vessel diameter and substance P is released from a nerve terminal to trigger a neurogenic inflammation (see Sect. 6.1.3). Many factors play a role in local pain processes. The substances released from tissue locally or from the nociceptor lead to vasodilatation, edema, and sensiti-zation of additional pain receptors through direct or reflectory factors. The following processes can be seen: 1 Muscle spasm. 2 Ischemic, metabolic, and physical injury stimulates nociceptive primary afferents which convey information centrally via the dorsal root ganglion to the dorsal horn. Degranulation of mast cells releases histamine. This stimulates nociceptors that release substance P, another potent vasodilator, to induce plasma extravasation. 3 The muscle spasm may be triggered, sus-tained, or aggravated by motor efferents from ventral horn cells. 4, 5 Sympathetic afferents and efferents monitor and influ-ence the diameter and tension of the vessel wall. 6 The vessel is shown first in a vasodilated state, later the diameter changes due to a vasospasm. Vasodilatation can be induced by sympathetic activity. Bradykinin, released from damaged tissue due to the muscle spasm, is also a potent vasodilator and increases capillary wall permeability to allow plasma extravasation. 7 The information from the primary afferent nociceptor is conveyed from the dorsal horn via the contralateral spinothalamic tract to the pain-processing areas in the brain. The intensity of the nociceptive information originating from the muscle can undergo significant modification before leaving the spinal cord

spine, the cervical spine has an especially large range of motion to all sides, which is made possible by the interaction of several vertebrae. In reclina-tion, the segments C2 to C4 are particularly in-volved, and in inclination, segments C1 to C5 (Fig-ures 90, 91, 92).

The vertebrae are linked by joint-like contacts, spinal joints, discs, and taut ligament structures with a good sensory supply, with the exception of the disc. The discs have great functional impor-tance. Through their elasticity and central plasticity, they promote the movement of the vertebrae against each other, in which displacement of the nucleus pulposus in the annulus fibrosus plays an important role. The gelatinous core glides forward when the spine is extended, the back of the disc becomes low-er and the front edge higher. In flexion, the process is reversed. This compression and release displace-ment moves the fluid contents of the cartilaginous fibrous tissue, ensuring that nutrients are distribut-

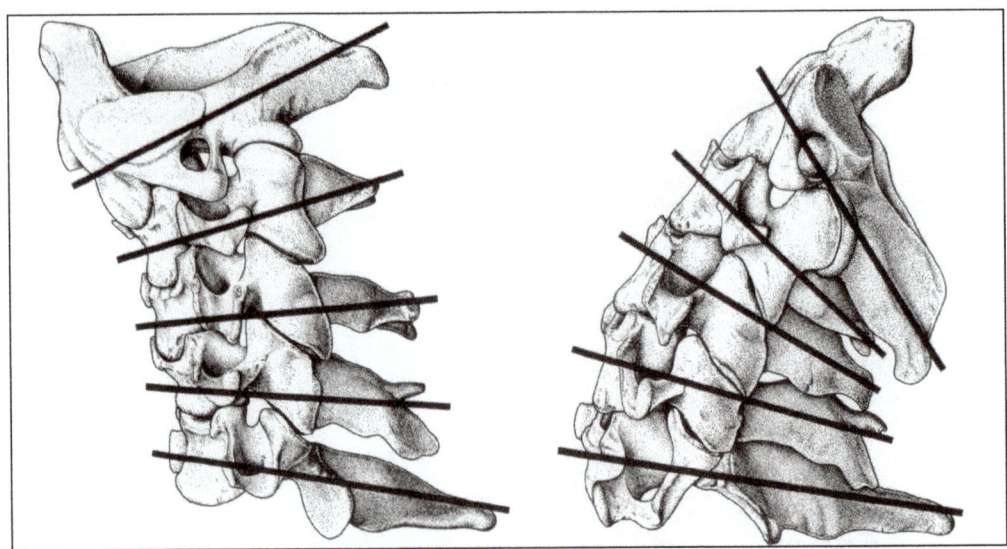

Figure 90 Mobility of the cervical spine. Inclination and reclination lead to pronounced changes of position of the vertebrae with respect to each other and stress the corresponding joints and ligaments

9

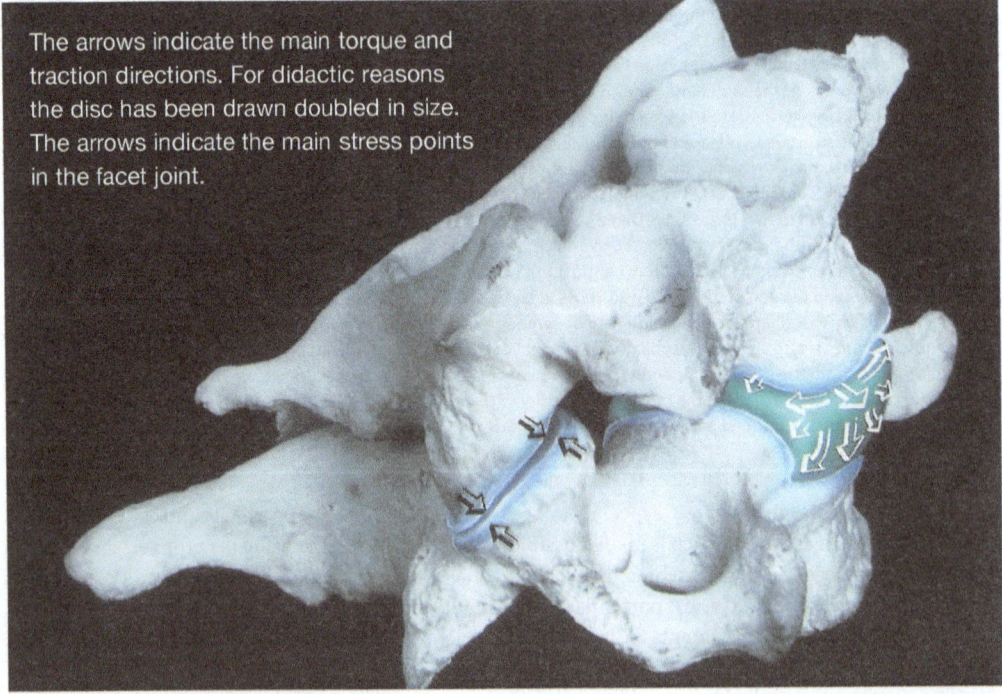

The arrows indicate the main torque and traction directions. For didactic reasons the disc has been drawn doubled in size. The arrows indicate the main stress points in the facet joint.

Figure 91 Intervertebral disc and facet joint

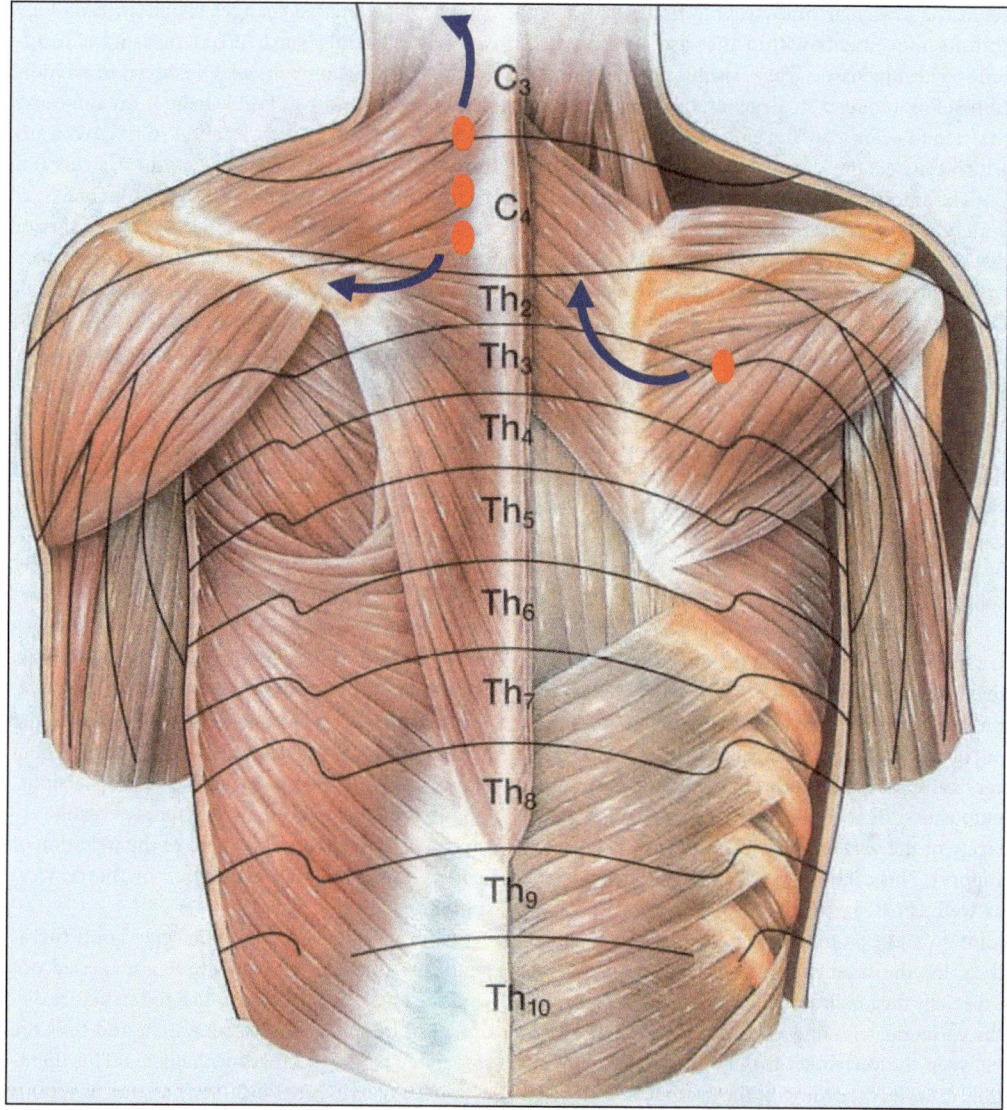

Figure 92 Spinal segments and muscles of the back. Scheme of the shoulder, neck, and back musculature. The reddish muscles represent various myotomes innervated from cervical segments and are layered over one another. Below the back muscles the thoracic myotomes are arranged in segments, clearly visible, and separated anatomically by the ribs. From the trigger points (red dots), pain can radiate in a cervical direction (see also Fig. 88). Phenomena of referred pain and pseudoradicular pain are difficult to distinguish from one another

ed in the avascular bradytrophic tissue. However, shifting movements within the disc are tolerated only by healthy tissue. The resistance in the annulus fibrosus is reduced in degenerative processes, so that small tears can develop during movement, which can become larger over time.

The anterior portion of the spine is composed of vertebrae that are separated by the intervertebral disc. The disc is a self-contained fluid elastic system that absorbs shock and allows fluid displacement. Due to this elasticity, distortion and movements between the bony segments are possible. On its traverse processes, the posterior portion has a joint with the next vertebra to permit and guide the movement of the adjacent vertebra. So the posterior articulation or "facet" forms joints that are lubricated with synovial fluid. They can undergo degenerative osteoarthritic joint changes. The joints are innervated by the articular branches of the posterior rami of the spinal nerve.

In addition to movement-related changes, there are other physiological fluctuations during the day that influence the thickness of the disc. For example, the static pressure when standing presses tissue fluid out of the disc – in the lumbar spine more than in the cervical spine. The padding function of the nucleus pulposus suffers from this loss of volume and the facets of the vertebral joints have to provide more support. This can lead to pain and a facet syndrome as well as osteoarthritic degeneration of the facet joint. In addition to affecting the bony segments of the spine, the degeneration also affects the ligaments, especially the longitudinal ligament, which connects the vertebrae with one another and fixates the discs between the vertebrae. If the discs are dehydrated, they exert less pressure on the ligaments. This relief that occurs during the day is compensated during the night. In particular in degenerative changes, the disc swells up and puts so much tension on the longitudinal ligament that the nociceptors in the ligaments are irritated and a resting pain occurs early in the morning. Such wear and reduction of the disc thickness is unavoidable over the years. The skeleton must accommodate the static displacement and degeneration. The extent to which this process elicits painful stimuli phenomena depends on the individual situation and the ability of the musculoskeletal system to compensate.

The degenerative changes visible in radiology cannot be generally considered a measure of insufficient ability to compensate. Of course, in an individual case, chondrosis visible in an x-ray can cause segment instability in a vertebral joint. Awkward movements then quickly trigger minor structural disorders, subluxation in joints, and stretched ligaments, which then lead to a reflectory restriction of movement via nociceptive stimuli. Since the entire paravertebral and back musculature is innervated, both sensory and motor, by the dorsal ramus of the spinal nerve, pain-reflectory processes can be easily clarified. Neck pain and "cervicogenic" headache, which is gaining recognition as an independent clinical picture, as well as pain in the shoulder and arm region, are understood as myofascial syndromes, and in the same sense brachialgia with sensorimotor deficits.

Irritation of the dorsal ramus can be caused by lateral herniated discs protruding into the intervertebral foramen or directly by degenerative changes in the joints of the cervical spine (Figure 96). This situation is discussed in more detail in Sect. 4, "Radicular lesions." Narrowing, similar to that in the intervertebral foramen, can also affect the foramina of the vertebral artery. There are controversial opinions on the issue of whether symptoms of vertigo are a result, and if so, to what extent, of the irritation of sensory and vegetative structures in the cervical spine and vertebral artery.

Chiropractic measures often give good therapeutic results. The manipulations are carried out with the idea that they penetrate fixed structure disorders in the musculoskeletal system and thus the associated pain-reflectory mechanisms. This therapy must be carried out by trained persons, as serious complications can occur, especially in the area of the cervical spine. As a precautionary measure, x-rays of the cervical spine must rule out fractures or metastases before treatment to prevent the risk of paraplegia from the manipulation. But even professional manipulation can create irritation of the vertebral artery that can lead to a vascular spasm, dissection, or thrombosis, from which serious damage from insults to the brain stem has been observed.

9.5.2 Whiplash trauma

The great range of movement of the cervical spine was pointed out in conjunction with the cervical syndrome. This range of movement can cause very serious damage if there is unchecked mechanical overuse. These situations are often observed in traffic accidents when the forward motion of a vehicle is abruptly stopped in a collision, or the reverse, if a stopped vehicle is suddenly accelerated by a crash from behind. If the passengers are caught by surprise in such a situation, they are not capable of fixating the head to the trunk using their neck muscles. Proper positioning of the headrest is thus an important way of avoiding injury. In an accident, various forces from opposite directions act on the head and spine. When a vehicle collides with a stationary vehicle in front of it, the car stops abruptly. But the head initially follows the original motion and flexes the neck. Then the head moves in the opposite direction, backward, and in doing so, hyperextends the spinal column, especially if there is no headrest. In the process, the centrifugal force is transmitted to the head and neck areas that brake it. The force is absorbed mainly by the osseous vertebral connections and ligaments. Only later do the shoulder, neck, and nape musculature participate reflectorily in the energy-absorbing maneuver. As a consequence of all these events, osseous injuries such as fractures or vertebral ring ruptures occur (Figure 93).

Luxations and overextension trauma cause tearing and hemorrhaging in the ligaments and capsules of the cervical region; contusions and strain can also extend beyond these structures to the musculature. The nerve roots and spinal cord often do not remain unaffected by the displacement of structures in the spinal column.

Pendulum movements of the head are performed more quickly by the bony skull than by the soft brain tissue embedded in the cerebrospinal fluid (Figure 94). As the latter only sluggishly follows the rapid skull movements, a biomechanical situation similar to spinal trauma arises. The movement vectors of vehicle and skull familiar from the accident scenario now play out between the skull and the brain. For this reason, symptoms of cerebral irritation and deficits are also observed in severe

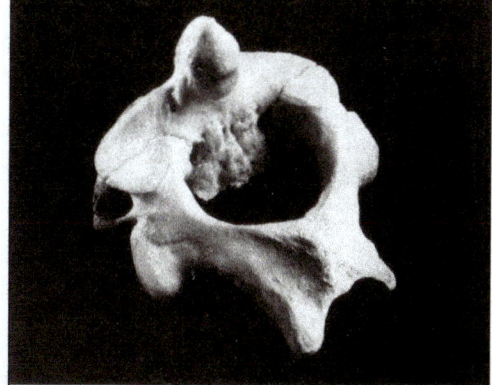
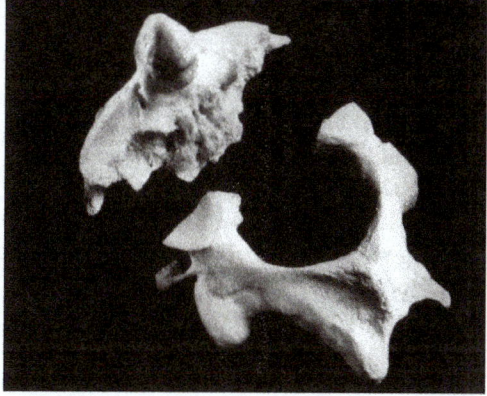

Figure 93 Cervical fracture at level C2. Due to a car accident, the anterior part of the C2 vertebra with the odontoid was fractured and dislocated from the posterior part

whiplash trauma. The following sequelae of trauma arranged by severity can occur:

1. Acute cervical syndrome: there are pseudo-radicular or occasionally clearly radicular lesions. The symptoms manifest immediately or within 12–48 h after the accident. Depending on the severity of the trauma affecting the cervical spine, the symptoms can last for days or weeks. The trauma with multiple tissue injuries is the ideal breeding ground for myofascial syndromes and their wide range of symptoms. It is still unclear to what extent vegetative reactions such as vertigo, nausea, sweating, flickering vision, and anomalies of the pupils can be explained by disorders of the nociceptive musculoskeletal system alone, or whether these symptoms must be interpreted as signs of cerebral damage.

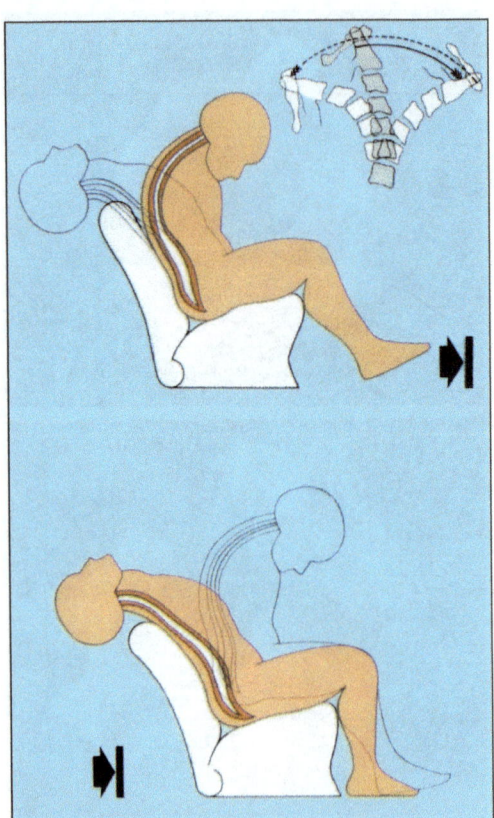

Figure 94 Mechanism of cervical spine injury. In whiplash trauma, damage to the spine and musculature is caused by hyperextension and hyperflexion of the cervical spine in the event of a frontal collision (A) ora rear-end collision (B)

2. Cerebral symptoms: irritation most frequently found in the frontal and temporal lobes after acceleration and deceleration trauma. As in commotio cerebri, which by definition never leaves permanent damage, nausea and short blackouts can occur. Irritation of the sense of balance can also be brought on by vertigo and hearing disorders. Among the substantial brain damage comparable with a contusio cerebri are permanent brain lesions with focal neurological deficits, persisting EEG changes, or neurophysiological disorders. Shear forces between the dura and brain can also bring about subdural hematomas.

3. Myelopathies: In addition to strains in the roots area with corresponding radicular

symptoms, paraplegic symptoms up to total paraplegia can occur.

As therapy, in overextension injuries, it is first useful to immobilize the joint connections with a cervical collar. However, it must not be forgotten that if immobilization lasts long, inactivity atrophy of the neck musculature can occur. Weak muscles tend to be insufficient when strained, which can develop into a cervical syndrome and thus lead to a vicious circle. This shows the need for the cervical collar to prevent involuntary wrong movements on the one hand, but parallel to that, the need for physiotherapy with specific muscle exercise and strengthening.

The extent to which psychological factors can influence the course of the illness was seen in Australia, when the insurance regulations, which initially quickly recognized and compensated for whiplash trauma, were toughened. This administrative action alone led to a drastic reduction of claims for possible accident-related whiplash trauma. This example makes it clear that the diagnosis of whiplash must be made very carefully and responsibly in order to avoid causing iatrogenic damage.

The average recovery time for all degrees of severity of whiplash is about 1 month. Only 12% of patients are still not in remission after 6 months. Factors prolonging the symptoms are: female gender, old age, tenderness and spontaneous painfulness of the neck musculature as well as numbness and pain radiating from there. Other factors for a worse prognosis are psychiatric illnesses in the past and negative sociodemographic and psychosocial factors of experiencing and handling illness. Up to a certain degree, chronic courses also represent therapy damage, caused by improper treatment, exaggerated fears, or inappropriate expectations. And last but not least, legal interventions are of considerable importance, not only for the time at which the case is concluded, but also for the duration of the symptoms.

9.5.3 Paravertebral pain

Cervical syndrome and lumbago are frequently associated with paravertebral pain that radiates locally into the musculature. The morphological, biochemical, and functional differences and chang-

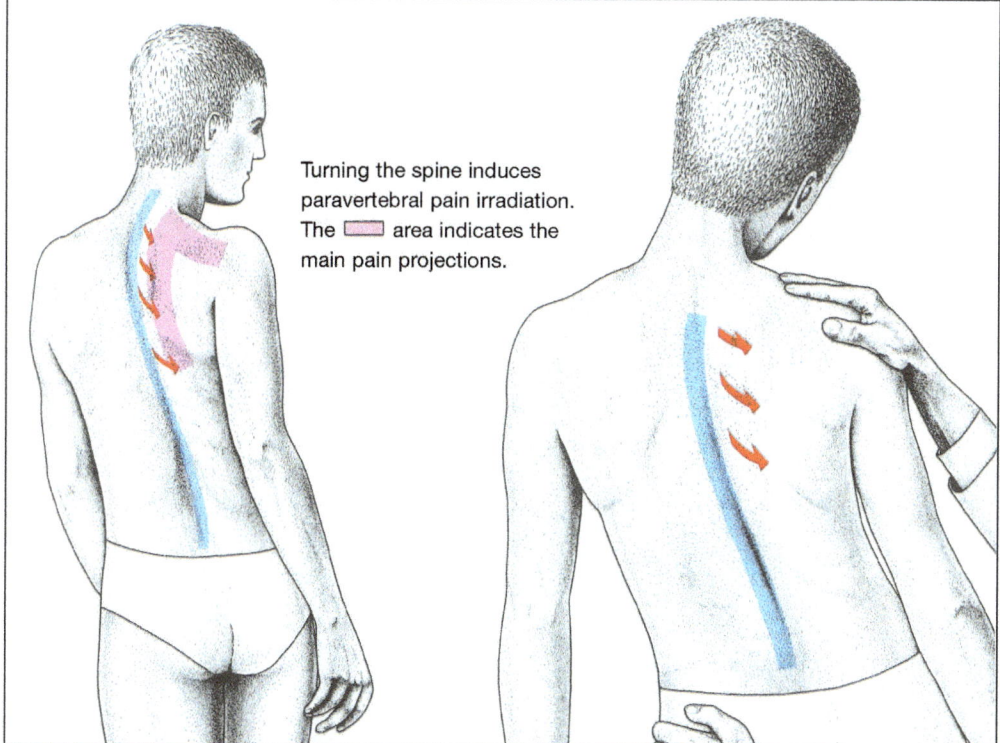

Turning the spine induces paravertebral pain irradiation. The [   ] area indicates the main pain projections.

Figure 95 Area of pain irradiation under lateral flexion of the thoracic spine

es that take place in acute and chronic diseases states have been explained in Sects. 4.2 and 10.3.8.

In addition to myogeloses, there is paravertebral pain on palpation in the area of the joints, the vertebral arches, and the ribs. If palpation yields an unremarkable situation, but there is movement-related pain when turning, bending, or stretching a spinal segment, the irritation is more likely to stem from the vertebral joints and their capsules and ligaments. Secondarily, there are almost always myofascial symptoms with myogeloses or tendomyoses. The pain state can have a simple explanation in structural disorders due to awkward posture. Trauma, degenerative joint disease, but metastases as well must be considered as potential causes and if they are suspected, must be ruled out with specific diagnostic measures. The dorsal branches are involved in pain perception and projection (Figures 95, 96), and their irritation can cause projection into the thorax wall with symptoms of intercostal neuralgia.

9.5.4 Low back pain, sciatica

The varieties of causative mechanisms described for the cervical syndrome are equally applicable to lumbago. Lumbago is one of the most common clinical symptoms seen by a general practitioner. From the medical and economic aspect, it has the features of widespread disease. Its clinical symptoms vary widely, ranging from severe, sudden onset of pain shooting locally into the lumbar region up to dull, difficult to localize pain radiating into the buttocks or legs. The symptoms can be acute or chronic. In the patient history, night pain is often mentioned, especially if the bed is too soft. The symptoms are localized in the lumbar muscles, where palpation shows a myogelosis. Myogeloses and trigger points are often located near the sacroiliac joint, which may play a role in pathogenesis. The range of movement of the lumbar spine can be severely limited and cause poor posture. Symptoms of musculature

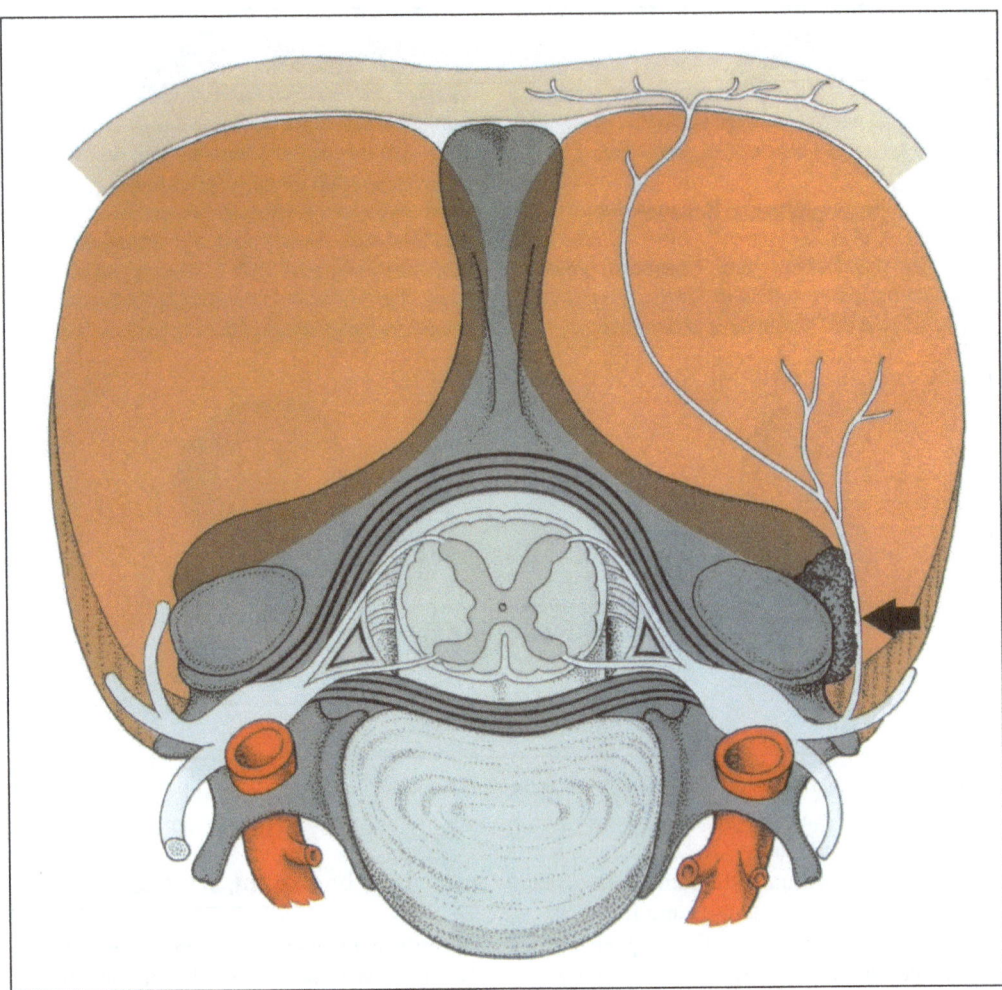

Figure 96 The dorsal ramus of the spinal nerve can cause back pain. Back pain can also be caused by irritations of the dorsal ramus. Frequently, degenerative changes occur in the proximity of the nerve, in particular of the vertebral joint facets, that elicit symptoms as in a mixed pain syndrome (arrow) and are perceived in projection in the back musculature or the skin

insufficiency with poor posture are just as often the cause of low back pain as instability in spinalpelvic posture and stress to the many ligament connections (Figure 98). Pain can also be a symptom accompanying urological, gynecological, or other disorders in the small pelvis (Figure 97).

Lumbago, sciatica, or low back pain are often casually diagnosed, all of which are diagnoses of a collection of various disorders (Table 16). Usually the first clinical signs of pain radiating to the lower extremities in patients with localized back pain are

interpreted as indicating co-involvement and injury to nerve roots. This is in particular the case when pain is associated with radicular sensitivity deficits, altered reflex patterns, and muscle paresis. In this situation, a discogenic cause of nerve compression such as intervertebral disc prolapse or sequestration must be ruled out. But very often, radiological examination shows that bony changes or postoperative fibrotic adhesions are the reason for the patient's complaints. If a disc prolapse is not released in time or nerve root compression persists for a

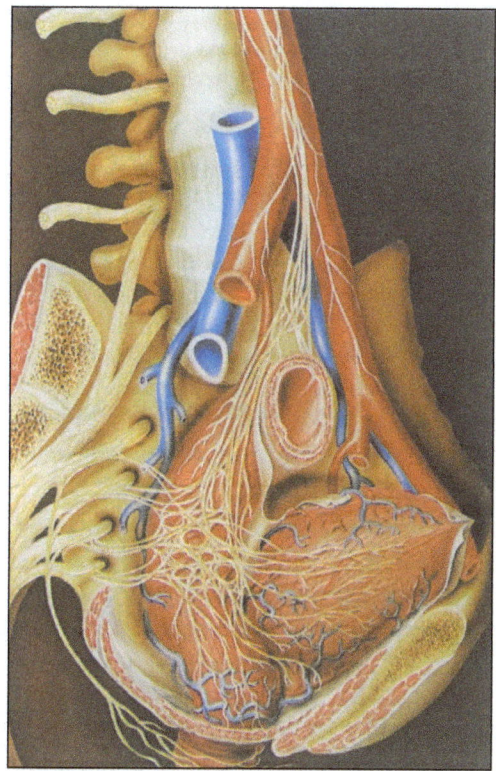

Figure 97 Autonomic nerve innervation and pelvis. In the small pelvis there is a dense network of somatic and vegetative nerve fibers supplying the pelvic organs

prolonged period, chronic nerve injury invariably develops. This often results in sciatic back pain with a neuropathic component. The combination of nociceptive and neuropathic pain mechanisms establish a mixed pain syndrome. The mechanisms creating such a syndrome are explained in Sect. 6.7 and Chapter 7.

Postural damage resulting in malpositioning of the small vertebral joints also leads, especially in the lumbar region, to cartilage damage and causes arthrosis or a facet syndrome. Despite this, disc damage is usually made responsible for the pain. This leads many patients to quickly come to the conclusion that they have disc or spinal damage and a disease history begins that not infrequently ends in unemployment or early retirement. With increasing age, radiological examinations of the spine show physiological degenerative changes and CT studies often show disc protrusions. But as long as there are no motor, sensory, or reflex deficits that can be localized to the radiology findings, the physician may not simply assume a correlation. The symptoms of radicular lesions are discussed in the relevant chapter of this book. The differential diagnosis must make it possible to distinguish between a lesion of the sciatic nerve, a symptomatic L5/S1 disc prolapse, or arthrogenic symptoms. This is significant, as lumbago often results from myofascial and ten-

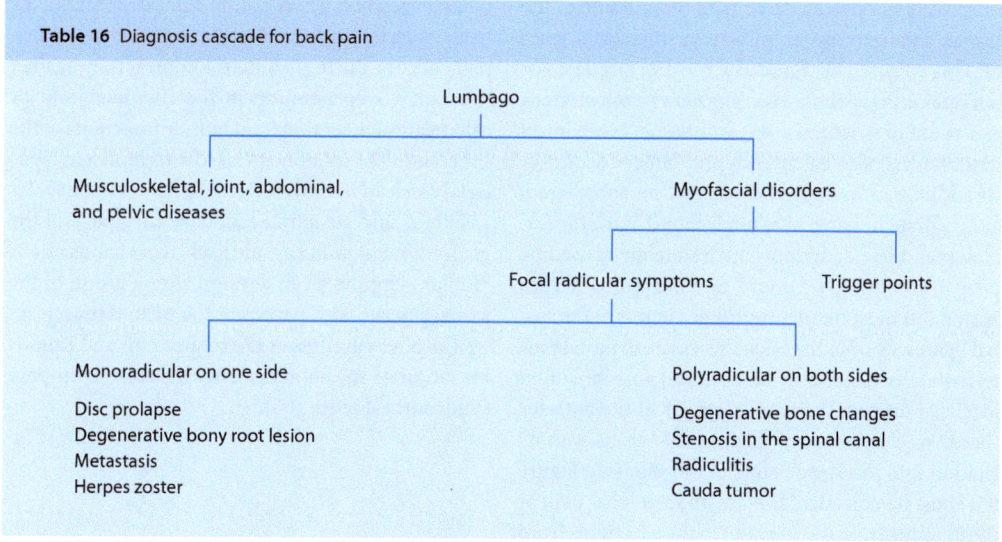

Table 16 Diagnosis cascade for back pain

Lumbago

Musculoskeletal, joint, abdominal, and pelvic diseases

Myofascial disorders

Focal radicular symptoms

Trigger points

Monoradicular on one side

Disc prolapse
Degenerative bony root lesion
Metastasis
Herpes zoster

Polyradicular on both sides

Degenerative bone changes
Stenosis in the spinal canal
Radiculitis
Cauda tumor

domyotic symptoms that require entirely different treatment than the radicular or arthrogenic injuries described above (Table 16). Especially for lumbago, therapy is dependent on the patient's own initiative. Effective prevention of recurrence must include an appropriate lifestyle and strengthening the musculature. Information on medication therapy is included in Sect. 10.3.2, "Non-steroidal anti-inflammatory drugs."

Hip pain, similar to sciatica irritation, can also occur in the course of degenerative processes, the most frequent being coxarthrosis with typical stress-related intensification of pain (Figure 98).

Hip movements restricted by pain can be determined rather easily while the patient is lying down by inner rotation and abduction of the leg. Vascular obstruction-related pain is also stress-related and can be recognized by auscultation and palpation of the vessels. Other considerations are insertion endopathies of the adductors and lesions in the sacroiliac ligaments, as well as hernias and inflammatory changes of the sacroiliac joint. Excruciating, burning pain at the coccyx after a sprain can lead to considerable discomfort as coccygodynia.

9.5.5   Intercostal neuralgia

Intercostal neuralgias are quite rare and are diagnosed too frequently in practice. An irritation of the intercostal nerves can cause pain on one side of the thorax, in one or more consecutive intercostal spaces. This is caused by trauma such as fractured ribs or operations in the thorax causing nerve contusions or as a result of strictures from scar tissue. Mechanical irritation can also occur with pronounced scoliosis. In addition, a prolapsed disc must be considered, although such changes are rarely found in the thorax. However, these symptoms are frequently caused by pleuritis, infiltrative tumors, or a beginning herpes zoster. But most frequently, trigger points at the spinal ligaments or in the back musculature imitate intercostal neuralgia. Since the painful areas are often localized only on the back, sides, or in the anterior thorax wall, a disease of the internal organs with irritation into the Head's zone of the skin (see Figure 86) must be considered. Point-shaped local pain at the juncture from the sternum to the ribs stems from

disorders of those joints or their cartilaginous connections. These symptoms are known as Tietze's syndrome and are shown in Figure 99.

9.5.6   Scapulocostal syndrome

This usually one-sided myofascial syndrome is often found in persons with work-related posture habits in which the shoulder girdle is overworked (secretaries, truck drivers). In arm and thorax movement, the muscles attached to the scapula are part of various myotactic units and are required for all movements. As already described for the cervical pain syndromes, cervical innervation offers various possibilities for projection-related pain radiating into the neck, shoulders, and arms including the hands.

9.5.7   Frozen shoulder

Shoulder lesions have a low tendency to heal spontaneously, and pain can thus be difficult to treat. With the exception of the acromioclavicular joint, the shoulder is supplied from the C5 segment. Different diseases cause identical pain that is not perceived in the shoulder itself, but at the upper arm, usually at the deltoid muscle insertion, radiating into the radial side of the forearm. If a movement is painful, the pain is caused either by tissue being stretched (non-contractile structures), or from muscles and tendons. This is the case when pain occurs during isometric contraction against resistance. Degeneration in the shoulder joint or calcification in tendons and muscle insertions of the short shoulder rotators (periarthritis humeroscapularis) can lead to movement disorders and muscular weakness and atrophy. Secondary shrinkage of the glenohumeral joint capsule leads to a vicious circle. Similar symptoms can develop after a lesion of the axillary nerve with paresis of the deltoid muscle, as well as after ruptures of the rotator cuff and immobilization of the shoulder after a stroke or in pain syndromes (Figure 100).

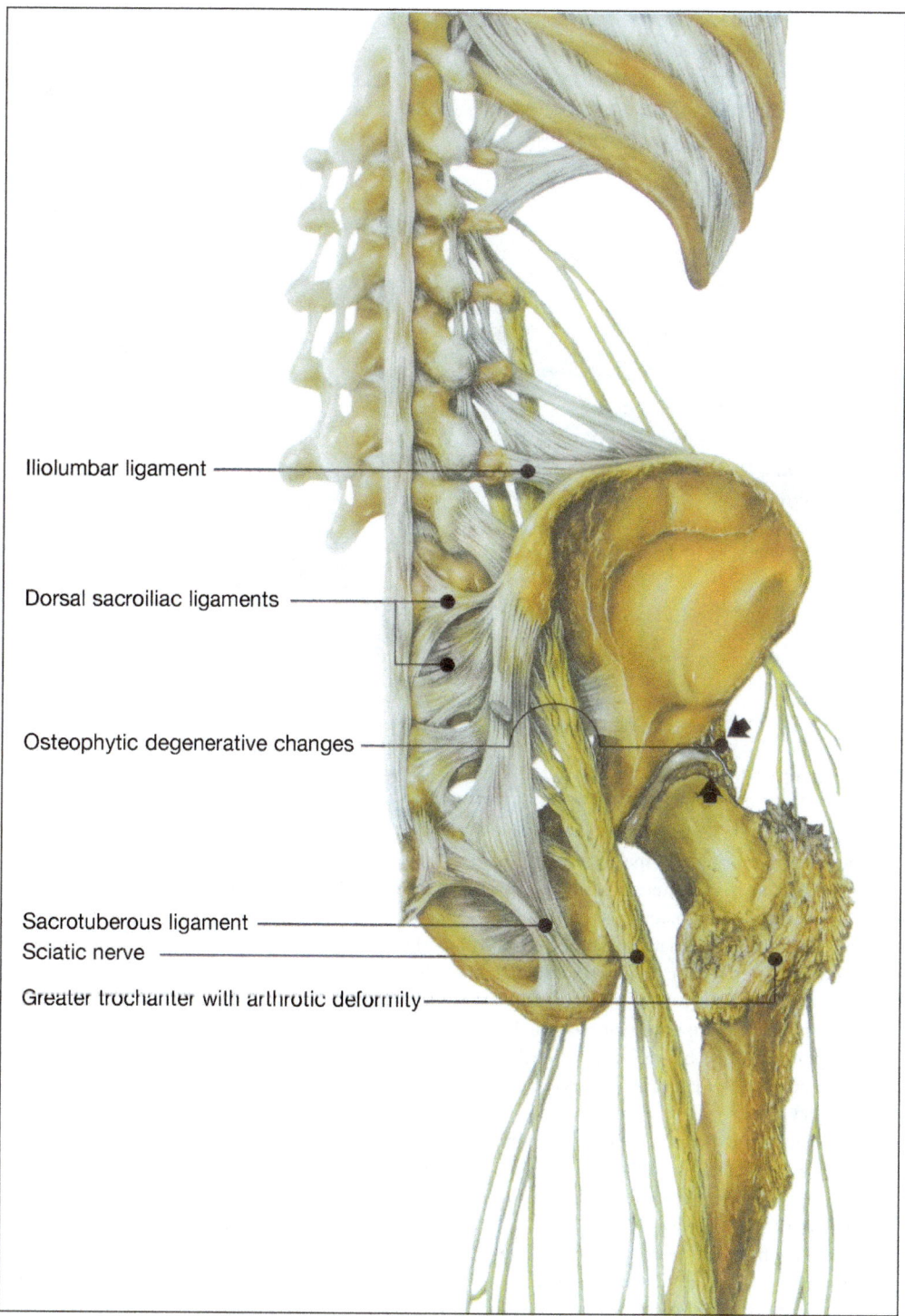

Iliolumbar ligament

Dorsal sacroiliac ligaments

Osteophytic degenerative changes

Sacrotuberous ligament
Sciatic nerve
Greater trochanter with arthrotic deformity

Figure 98 Osteoarthritis, coxarthrosis. Coxarthrosis and other degenerative changes of the bones and ligaments in the pelvic region often cause hip and back pain

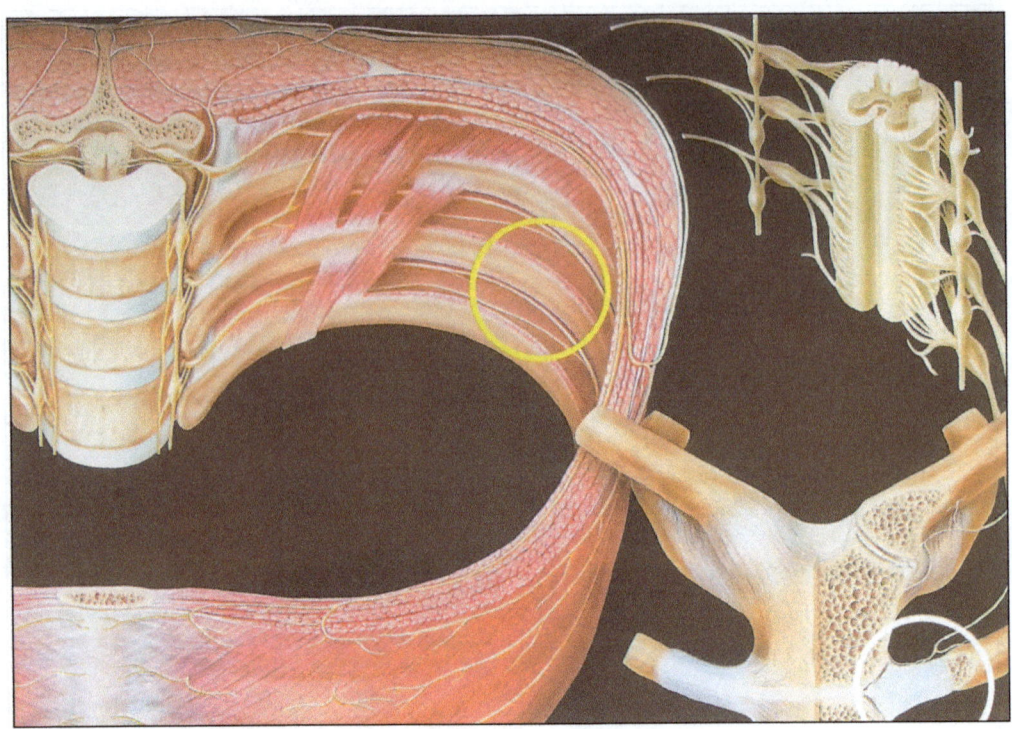

9.5.8 Brachialgia

Factors triggering brachialgia are:
- Degenerative changes to the cervical spine with musculoskeletal and referred pain.
- Periarthritis humeroscapularis.
- In addition to pain radiating from the cervical region, pain can also be projected from the hand upwards. The carpal tunnel syndrome is the most important peripheral neurological disease.
- Symptoms can also be triggered by mechanical, inflammatory, or malignant processes in the brachial plexus.
- Non-neurological causes are pain from improper use and overuse as well as circulatory problems occurring from arterial stenoses or axillary vein thromboses.
- Epicondylitis causes tenderness at the origin of the long hand and finger flexors at the lateral elbow. The typical maximum pain pressure point is in the region of the radial head. Diffe-

rential diagnosis: lesion of the posterior interosseous nerve of the radial nerve.

9.6 Herpes zoster

Herpes zoster, also known as shingles, is triggered by the varicella zoster virus. The first contact causes chicken pox, the second herpes zoster. Herpes zoster results from reactivation of dormant neurotropic varicella-zoster virus in a sensory ganglion and is usually manifested as an acutely painful vesicular rash affecting a single dermatome, which generally resolves within a few weeks. There is no correlation with herpes simplex. About 4 out of every 1,000 people contract herpes zoster, but 30% of the cases are inapparent. Zoster infection affects the ganglia of the cranial and spinal nerves with inflammatory and hemorrhagic changes. At first there is viral replication causing ganglionitis, and then the virus spreads along the nerve to the skin. Since the in-

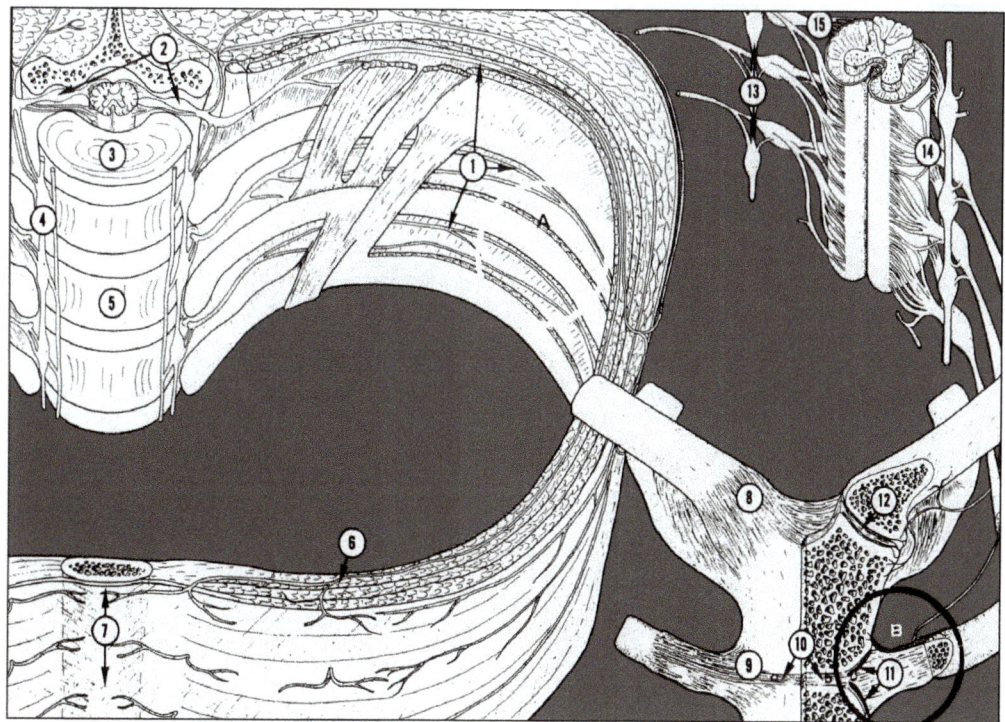

Figure 99 Anatomy of the chest and its innervation. A Intercostal neuralgia, 1 Nervi intercostales, 2 Ganglion spinale, 3 Discus intervertebralis, 4 Ganglion trunci sympathici, 5 Corpus vertrebrae, 6 Ramus curaneus anterior n. intercost, 7 Sternum, 8 Tietze's syndrome, 9 Ligamentum sternocostoclaviculare, 10 Ligamentum sternocostale intra-articulare, 11 Sternocostal joint, 12 Discus articularis, 13 Sympathetic trunk, 14 Radix ventralis, 15 Radix dorsalis

flammation can often extend to the spinal cord and dorsal horn (Figure 101), the cerebrospinal fluid shows inflammatory changes with an elevated cell count. The infection is most common in the thoracic area, followed by the trigeminal nerve and the ophthalmic branch, and then in the cervical or lumbar region. The first symptoms are often tingling, itching, and the patient scratches the skin because of pruritus. Then the rash begins as macules and papules, which evolve into vesicles and then pustules. Generally the disease resolves within a few weeks, and only a minority of patients suffer from postherpetic neuralgia.

Risk factors for complications and postherpetic neuralgia are severity of the acute infection, advancing age, and reduced immune status as well as sensory abnormalities such as hypoesthesia and allodynia. Motor paresis can occur 1–2 weeks after the initial pain symptoms and is often overlooked as a result of the usual adaptive posture that is almost always adopted due to pain. Concomitant myelitis and encephalitis or generalization of the pain is rare. Generally, one or more cranial nerves or dorsal roots are affected. In cases with only slight skin eruptions, the radicular, sensory, and motor damage can be confused with a prolapsed disc. A thorough physical examination with inspection of the skin avoids this diagnostic trap. In the cranial nerve region, there may be inflammation of the trigeminal nerve and the risk of conjunctival and corneal damage with zoster ophthalmicus. In rare cases, zoster angiitis accompanying zoster ophthalmicus can be the cause of a cerebral insult. Facial nerve paresis often occurs after zoster oticus, which is marked by burning pain in the auditory canal. Changes in tear secretion and sense of taste indicate an inflamma-

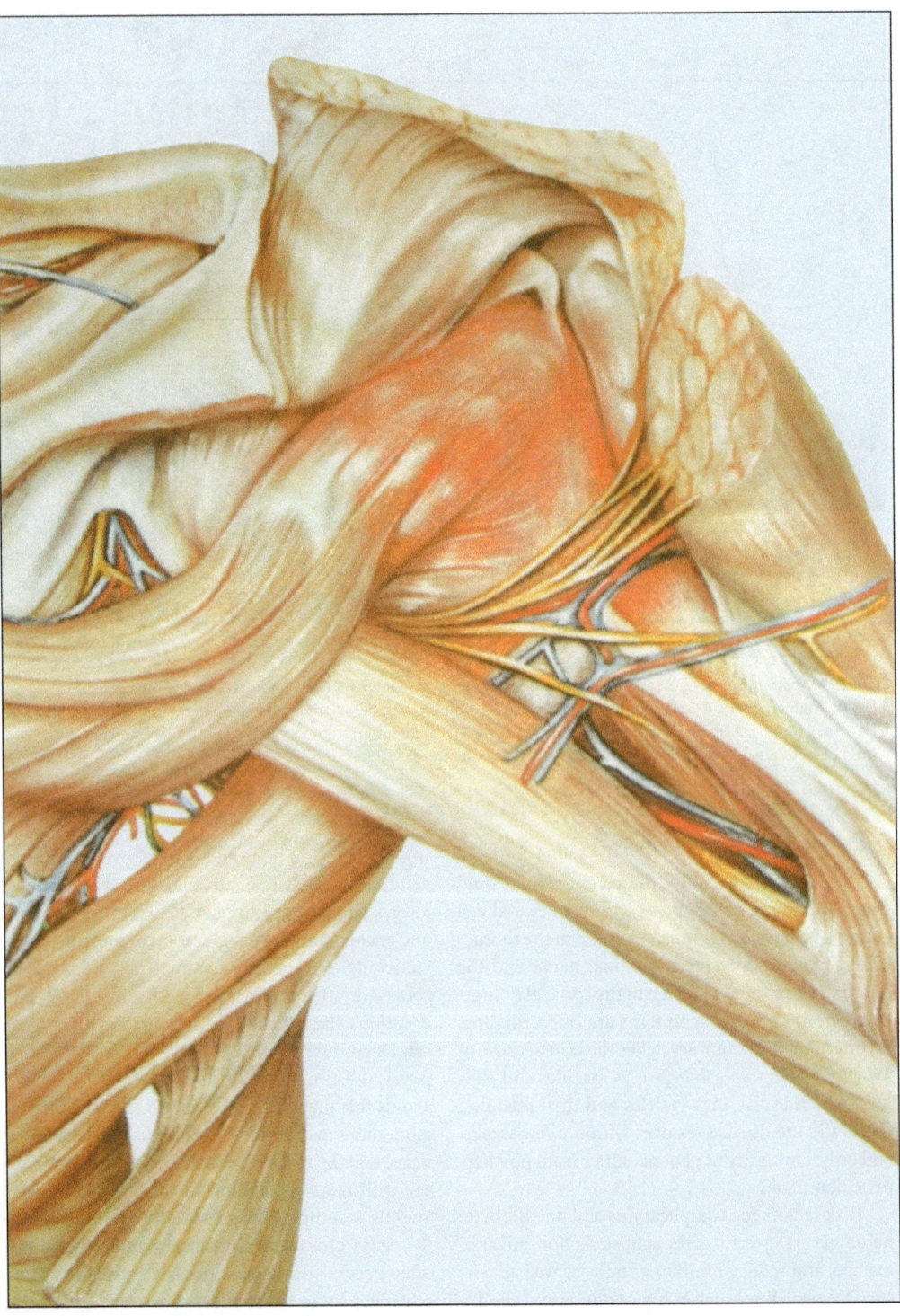

Figure 100 Inflammatory arthritis as one reason for a frozen shoulder

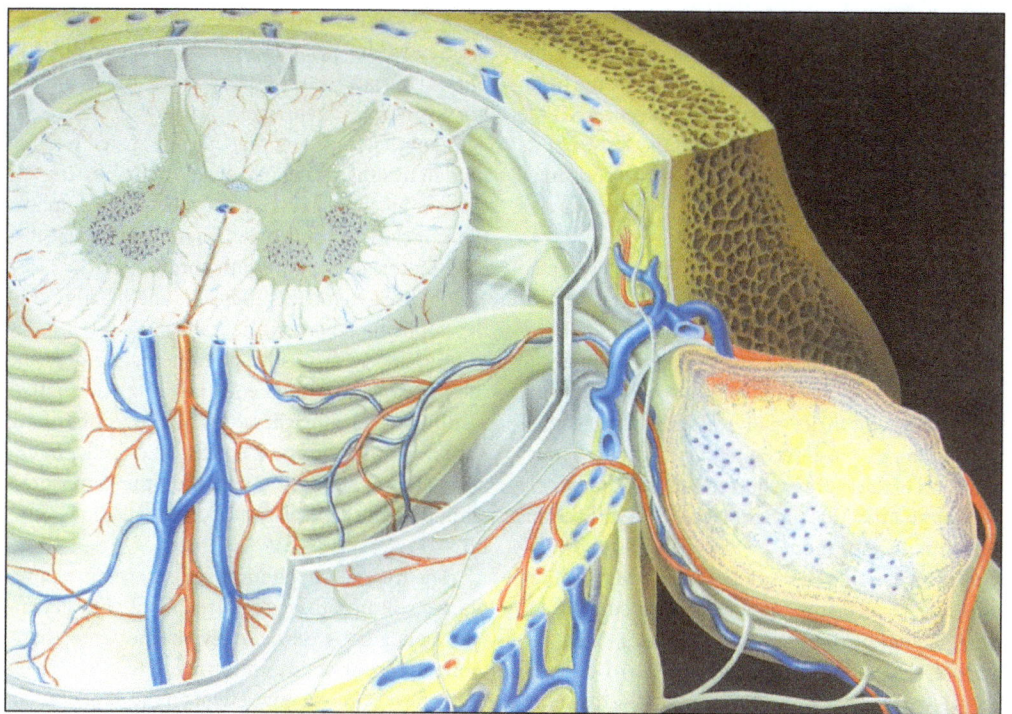

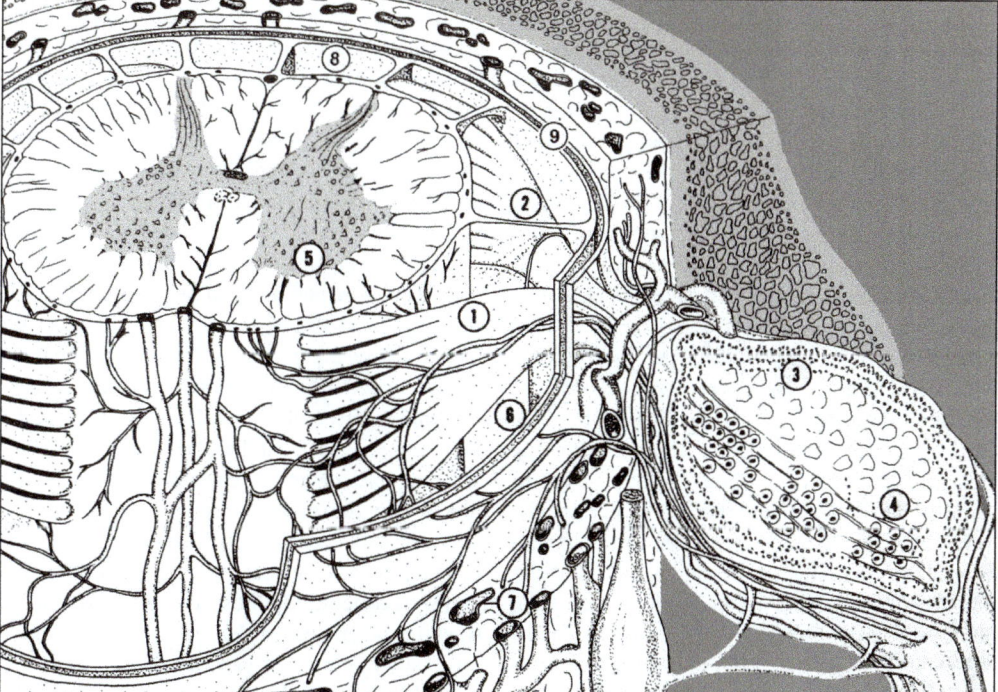

Figure 101 Acute herpes zoster infection. 1 Ventral root, 2 Dorsal root, 3 Marginal bleeding in the spinal ganglion, 4 Lymphocyte infiltrates and lipid phagocytes, 5 Ventral horn, 6 Subarachnoid space, 7 Venous plexus, 8 Subdural space, 9 Epidural space

tion in the region of the geniculate ganglion in the facial nerve canal. This diagnosis is made under the syndrome known as Ramsay Hunt syndrome by an inspection of the ear and the efflorescences visible there. Zoster usually occurs spontaneously, but its appearance is also often associated with consumptive diseases as well as immunosuppressive illnesses and therapy.

In the acute stage, zoster can be easily treated with the guanosine analogues acyclovir, the prodrug of valacyclovir, and famciclovir. The oral bioavailability and levels of antiviral drug activity in the blood are higher and more consistent in patients receiving thrice-daily valacyclovir or famciclovir than in those receiving acyclovir five times daily. All of these drugs shorten the duration of skin lesions and the incidence of visceral complications in immunocompromised hosts. The issue of dosage, as well as of oral or parenteral administration, must be decided from case to case. In comparison with biologically well-reacting persons, immunosuppressed patients have a higher risk of developing severe and generalized variants of the disease. If treatment with guanosine analogues is initiated within 72 h of the appearance of efflorescences, it will be successful and is also considered to have a favorable influence on the frequency of unwanted secondary symptoms. Their effectiveness in preventing postherpetic neuralgia has not yet been proven. In the beginning acyclovir should be given 400–800 mg orally five times daily for 7–10 days. For severe forms and high-risk patients, an infusion of 10 mg/kg body weight three times daily should be given immediately for 7–10 days. Famciclovir is given 500 mg orally three times daily for 7 days; valacyclovir is given 1 g orally three times daily for 7 days. In addition to analgesics, corticoids can also be administered for pain. Prednisone can be given 60–80 mg/day for 10 days, at a gradually decreasing dosage. Glucocorticoids should only be given in combination with antiviral therapy. In all cases, for severe pain, good results have been achieved with topical local anesthetics or capsaicin 0.025%. It should only be applied to intact skin. As capsaicin initially irritates the nociceptors locally, care must be taken to ensure that it does not come into contact with the eyes or mucosa outside the affected area.

A live attenuated herpes zoster vaccine is recommended for persons 50 years of age or older to prevent herpes zoster and its complications, including postherpetic neuralgia.

Experience shows that herpes zoster is often mild in healthy young persons. With increasing age there is an increased risk for pain and complications, including postherpetic neuralgia, ocular disease, motor neuropathy with paresis, and CNS vasculitis disease. All this increases the likelihood of developing postherpetic neuralgia.

Patients with postherpetic neuralgia describe their pain as a spontaneous ongoing pain with a continuous burning character. Moreover, they have paroxysmal shooting or electric shock-like pains, and allodynia to light touch and other innocuous stimuli or mechanical hyperalgesia to noxious stimuli.

The treatment options for postherpetic neuralgia are still unsatisfactory. Treatment with guanosine analogues is very effective in the acute phase, while glucocorticoids are of restricted use. These drugs have no preventive effect on the development of chronic postherpetic neuralgia. On the basis of strong evidence, treatment guidelines for postherpetic neuralgia recommend tricyclic antidepressants, serotonin re-uptake inhibitors, and gabapentinoids. Amitriptyline should be started with dosages of 10–25 mg and increased in small increments of 10–25 mg weekly up to 150 mg/ day. If this regimen does not relieve the pain, carbamazepine, gabapentin (starting with 100–300 mg at bedtime, max. daily dose 3,600 mg), or pregabalin (75 mg twice daily, max. daily dose 600 mg) can be given in addition. All drugs must also be started at low dosages with increasing weekly increments and maximum daily doses divided into two or three doses. As second-line agents, topical capsaicin 0.075% or local 5 % lidocaine patches can be used. Tramadol is used at 50 mg once or twice a day and can be increased up to 300 mg, but its use is controversial. A Cochrane review in 2013 concluded that there was not convincing, unbiased evidence of a benefit of oxycodone treatment. Opioids, including tramadol, should generally be considered as third-line drugs for postherpetic neuralgia because of safety concerns. It should only be used after consultation with a specialist and should be prescribed only with appropriate goals and close monitoring. Nonsteroidal anti-inflammatory drugs are generally considered to be ineffective for neuropathic pain.

Inconclusive effects have been reported for cannabinoids, valproate, topical clonidine, and antiepileptics. There is no convincing evidence for treatments with local anesthetics or nerve block of the sympathetic nervous system.

To tailor therapy, quantitative sensory testing (QST) can be helpful. Patients with hyperalgesia will likely profit from oxcarbazepine since upregulated sodium channels are responsible for ectopic activity in the damaged neural structures.

9.7 Muscle cramps

Nearly everybody will eventually experience a sustained painful contraction in a whole muscle, parts of a muscle, or a muscle group. Muscle cramps are common incidents. Cramps are uncommon in children but with increasing rate they are a nuisance in people older than 65 years. Mostly they occur in the calves and at night. Cramps at night will interfere with sleep and with quality of life. These cramps are self-limiting but can be followed by muscle soreness. Physiological conditions that trigger cramps are strenuous exercise or a period of inactivity like night rest.

Many hypotheses have been put forward about the etiology of this condition and its predisposing factors. Among the physiological conditions is exposure to heat alone or in combination with dehydration as well as electrolyte imbalance such as hypomagnesemia. Lifestyle should be discussed during medical history-taking, as well as present medication for and possible indicators of neurological diseases.

Neurological diseases that are associated with cramps are amyotrophic lateral sclerosis, myopathies, and hereditary neuropathies such Charcot-Marie-Tooth neuropathy. All polyneuropathies can be accompanied by muscle cramps. Diabetic polyneuropathy comprises the largest group of sufferers. They can have cramps not only in the calves but also in the proximal legs, trunk, and upper extremities. As diabetic patients very often have vasculopathies, the vascular status should be examined for vascular claudication. Ischemic situations can mimic or trigger cramps. Liver disease and renal dysfunction can influence homeostasis of amino acids, chloride, and potassium. Metabolic derangement may also be due to parathyroid dysfunction, malnutrition, or malabsorption. This can cause electrolyte deficiencies for magnesium, calcium, and potassium as well as vitamin B or D hypovitaminosis. Medications that have been reported to produce cramps as side effects are statins, thiazide diuretics, and acetylcholinesterase inhibitors.

Cramp fasciculation syndrome is a rare clinical entity. The clinical picture consists of cramps and isolated fasciculations. A fasciculation is the result of spontaneous firing of a motor unit. All muscle fibers belonging to this unit contract simultaneously. Uncoordinated contractions of single muscle fibers form the picture of myokymia. Fasciculations can often be seen especially in the calves. Because this muscle twitching can be intensely annoying and disruptive, patients are often worried. Since fasciculations are also seen as the first sing of a motor neuron disease, it raises additional anxiety. Careful neurological examination can reassure patients that it is a benign but painful syndrome without weakness and atrophy. Cramp fasciculation syndrome is idiopathic and believed to be a hyperexcitability syndrome of the peripheral nerve. Carbamazepine and phenytoin have been reported to decrease the signs of the nerve hyperexcitability and to reduce cramps.

The reason for cramps can be high-frequency firing of motor neurons due to triggers in intra-muscular nerve terminals, along the motor axon or neuron. Abnormal excitability can be the result of changes in electrolyte concentration around the motor endplate or mechanical due to tendon shortening in old age or inactivity. As in neuropathic pain, the motor nerve can be injured and ephaptic transmission as well as sensitization can occur. At the spinal level, the motor system is connected to sensory input.

To prevent muscle cramps, stretching of the muscle can help prior to exercise or after prolonged inactivity. There is no medication that can definitively prevent the occurrence of camps. Therefore there are many reports of different drugs that may be helpful. Quinine sulfate at doses ranging from

150 to 450 mg/day has been reported. Owing to id-iosyncratic reactions and reported deaths, the use of quinine is forbidden in the USA. Other reported medications are the use of calcium channel blockers such as 30 mg of diltiazem hydrochloride or low-dose baclofen. A treatment with vitamin B complex and 30 mg vitamin B_6 was also reported to be help-ful.

9

General therapeutic guidelines

There are currently no methods available that have been proven to promote or accelerate regenerative processes in nerves. This lack often leads to disappointment for both patient and physician. It is therefore not surprising that this therapeutic vacuum gives rise to a number of pharmacological, technical, and other rehabilitative measures of varying quality for bridging the difficult time between a nerve lesion and the final condition. In this situation, therapy becomes an art: when scientific know-how and empirical experience are coordinated to the needs of the patient to determine further procedure.

The inability to directly influence nerve regeneration adds extra weight to conservative treatment measures, which, as far as can be estimated today, can create favorable conditions for the regeneration process and the functional result of regeneration. These different possibilities and measures, whose uses are presented here, are required for all degrees, variations, and causes of peripheral nerve damage. It is not particularly useful to address individual diseases, as the same basic principles are always applicable.

10.1   Nerve-muscle interaction

Therapy approaches must always take account of potential sequelae and their dynamics. The most obvious are muscular changes with the initial limp paresis and the atrophy that follows after a few weeks. The degeneration of the peripheral axons leads early on to inexcitability and blocking of impulse transmission and secondarily to destruction of end plate formations.

The functional unit of nerve and muscle is established in the course of development. The previously undifferentiated musculature is differentiated when innervated under the functional effect of the nerve into muscle fiber types with different contractile, metabolic, and structural characteristics and membrane properties. Presumed to be mediators of this development are a trophic factor transmitted from the nerve to the muscle on the one hand, and on the other, the discharge pattern of the motoneuron. The trophic factor is assumed to be a chemical substance that passes over from the nerve to the muscle at the neuromuscular synapse and controls the development of its genetic pattern. If a rapidly contracting muscle is exposed to a slow excitation pattern, it is transformed to a slow muscle, and the reverse form of influence is also effective. Since the denervated muscle also reacts in this manner, the excitation pattern must be of critical significance. The muscle cell does not display denervation behavior until the axon terminals and, with them, the neuromuscular end plates have degenerated.

Within the axon, neurotransmitters, proteins, phospholipids, etc. are transported in a fast and a slow fraction in proximo-distal direction. The sequence of the accompanying phenomena of this wallerian degeneration follows the flow velocity of axoplasmic transport. The interval up to the start of denervation of the muscle depends on the length of the distal axon segment. The recoupling from the periphery to the center (cell nucleus) is attributed to an additional retrograde axoplasmic flow. Accordingly, the start of chromatolysis seems to be dependent on the distance between the cell soma and the lesion site on the axon. At the same time, a partial separation of the cell from the synaptic bond occurs, which is confirmed physiologically by the reduced reflectory excitability and anatomically by the displacement of the synaptic end-feet.

10.1.1 Electrotherapy of the muscle

Initially conducted and propagated by Reid (1841), and broadly used on the many nerve injuries that were caused in the world wars, the results of the electrical stimulation of denervated muscles under therapeutic aspects were still so contradictory that the benefits of electrotherapy are assessed in various ways to this date.

Through the loss of the trophic influence of the nerve, typical degeneration occurs in the muscle and it atrophies. An attempt is made to use electrical stimulation to halt this degenerative process in the muscle and delay atrophy. In animal experiments, an effect of delayed muscular atrophy was achieved by appropriate stimulation processes. However, this success is linked to firmly defined conditions and is endangered by even a slight modification. The conditions listed in Table 17 must be maintained as the basic conditions for successful electrotherapy.

Table 17 Standards for electrotherapy of denervated muscle

1 The musculature must contract repeatedly under electrical stimulation.

2 Contraction must take place with good strength development with no significant shortening of the muscle (isometric).

3 Electrical muscle stimulation with abundant contractions must be carried out for at least 20 min/day.

4 Breaks are required, as the muscle fatigues quickly.

5 Treatment must be given every day.

6 Tetanic muscle contractions must be avoided, as they promote the formation of fibrotic tissue in the muscle.

Muscle stretching as a growth stimulant

In animal experiments, it was shown that even unstimulated, merely passively stretched muscles atrophied more slowly. The stretching impulses developed in isometric contraction appear to delay muscle atrophy. Stretching impulses trigger the re-synthesis of the substances in the muscle that are otherwise rapidly degraded due to degeneration. Thus, a balance between synthesis and degrading can be maintained over a long period; as a consequence, stimulation delays atrophy. If, however, no stretching forces can be brought about in the muscle, therapy is less useful. This situation always exists for facial muscles. There is no fixed insertion corresponding to the fixed origin. The musculature in the face cannot be exercised isometrically, and electrical stimulation of these muscles is difficult to dose. Every stimulus triggers clearly isotonic muscle contractions. If it is too strong, the musculature can be surrounded by connective tissue and the feared facial muscle contractions develop.

Experimental findings are not a recipe for practical application

The successes shown in animal experiments cannot be simply transferred to humans. There are special conditions in animal testing. First, only there is intensive and regular therapy guaranteed. Secondly,

the animal muscles to be treated are relatively small, and thirdly, the stimulus electrodes used in treatment occupy disproportionately large areas. Only if the entire muscle is sufficiently perfused can all of the muscle fibers within a muscle be made to contract. Therapy is thus useful only under these conditions. Human musculature has a considerable larger volume and requires greater stimuli for perfusion. But these stimulus strengths cannot be applied to human muscles, as they are associated with electrical currents that are perceived as extremely painful. Any painful stimulus must be avoided at all costs. It is known that pain stimuli can change the circulation of denervated extremities. In addition, reduced perfusion promotes the formation of connective tissue in the muscle.

In the electrotherapy of denervated musculature in humans, the issue must be addressed as to how long electrotherapy is useful. This question can be answered only in conjunction with the severity of the nerve lesion. For long regeneration distances, the reinnervation of a muscle can take many months. In animal tests, it has not yet been proven whether electro-stimulation actually delays atrophy over many months. However, another aspect is also of great significance. Animal tests show that even after several weeks of denervation, the denervated musculature recovers very quickly, as soon as it is reinnervated. And reinnervated muscles were no different from the previously electrotherapied muscles after a few weeks. This means first that as soon as the nerve can again exert its trophic influence on the muscle, the muscle recovers quickly, and second, that the atrophydelaying effect of electrostimulation offers no functional advantage for a treated muscle over an untreated muscle.

10.1.2 Electrotherapy of the regenerating nerve

Under the aspect just presented, the idea suggests itself that the difficult to dose and hardly measurable voluntary activation could be replaced by artificial activation. Indeed, electrical stimulation of the damaged axon brought about an increase in protein synthesis and an acceleration of axoplasmic flow. The potential effect appears to be worthy of clinical testing due to observable changes of the axoplasmic flow in polyneuropathies. The influence promoting the regeneration of bruised axons could be achieved by low-frequency stimulation of the nerve proximal to the lesion site.

It remains for other studies to integrate such aspects in a clinically justifiable therapy program through a broad confirmation of the individual aspects existing up to now.

10.1.3 Physical therapy

Caused by denervation, changes occur in muscles and joints of the affected extremity that require physical therapy. The individual muscle fibers lose sarcoplasm at a rate of about 50% in 4–6 weeks. A corresponding loss of volume and weight can be determined. The percentage of connective tissue initially remains constant, resulting in a relative increase in connective tissue in the endo- and perimysium. As the period of denervation increases, the amount of connective tissue also increases absolutely. No significant structural changes occur in the vascular bed of the muscle. The initially disrupted circulation while resting later returns to normal levels, but retains its lowered adaptability. It is likely that a circulation-related reduced supply of the muscle cells promotes atrophy. Venous congestion of the muscle that can occur from gravity alone as a result of lower vascular tone and the absence of muscle contractions, especially in a hanging extremity, also exert a negative influence. Congestion and reduced lymph drainage stimulate the proliferation of connective tissue.

These changes can be countered by appropriate positioning of the affected extremity and careful stroking massage in the direction of the trunk. Painful manipulations are to be strictly avoided. But improvement of muscle strength or circulation are not to be expected from this therapy.

Immobilization and sensory deficits eventually lead to secondary changes in the musculoskeletal system. Therefore, a suitable position of the extremity in a functionally favorable "medium position" in the event of an irreversible joint fixation must be attempted. Joint positions that allow a complete "unstretching" of the paralyzed muscle lead to contractures and permanent shortening of the muscle from interstitial fibrosis. Conversely, if there is long-term overextension of the paralyzed muscles, a marked delay in the recovery phase can be expected. In positioning, it should also be ensured that the joints can be moved passively by physical therapy and actively by the intact musculature. If a joint is moved within its physiological range of movement, this has a favorable effect on the ligaments, tendons, muscles, and connective tissue around the joint. Movement stretches the structures so that they retain their physiological function and especially so they do not shrink. Such joint stiffness from a lack of exercise also occurs if there are inconspicuous innervation conditions, but the patient is immobilized due to fractures or central damage such as after a stroke. The frozen shoulder syndrome can then appear at the shoulder (see Sect. 9.5.7).

Sensory disorders lead to the risk of pressure-related lesions on the skin. The proper precautions must be met.

Heat application is a problematic measure. Neurogenic disorders of circulatory regulation combined with reduced sensitivity to pain can cause thermal damage. Even when applied physiologically, heat application promotes the formation of edema. Already existing edema can be increased by heat and is thus a contraindication for heat therapy because the fibrotization of the muscle would be accelerated.

10.1.4 Voluntary activation

Paretic muscle groups can be disconnected from the movement pattern. A reafference disorder possibly plays a significant role in this. Clinical observations show that nerves that are under continual physiological central stimulation undergo a more rapid functional regeneration than nerves that are not exposed to corresponding activation.

Such a physiological activation can sometimes be achieved with tricks. If applicable, an attempt should be made using electrical stimulation of the denervated musculature or by passive movement to retrain the lost feeling of movement in order to reintegrate the paralyzed segment into the movement pattern and thus prepare it for voluntary activation.

Other options are targeted training of an intact synergist of the paralyzed musculature, by which means co-activation of the neuron population for the affected musculature can be achieved, even if the effect is not initially noticeable. Less effective is an activation attempt using strong voluntary tension of symmetric muscles on the intact contralateral side. This principle of "cross-education" is made possible by the residual infantile bilateral innervation scheme.

Myofeedback

Long before the beginning reinnervation of perceptible, the possibility of voluntary reinnervation can be determined by EMG. From this time on it is possible to make the not otherwise recognizable activity optically or acoustically perceptible to the patient. This is done using EMG devices by playing the signals on a monitor or over loudspeakers. This procedure not only increases motivation for the patient, but also makes it possible to practice reafference. The procedure is therefore used regularly in therapy at many rehabilitation centers.

10.1.5 Motivating the patient

All therapy measures are useful only if they are implemented as early as possible. In addition to all the measures mentioned, it is especially important to motivate the patient He should participate actively and undergo all the various procedures daily and, in particular, do innervation exercises repeatedly on his own. If the patient is not motivated positively, the results are much poorer than otherwise possible.

10.2 Regeneration-promoting substances

The direction of therapeutic considerations is determined by the respective injury. For example, at the start of every polyneuropathy therapy is the search for and neutralization of the noxious cause, such as alcohol, a diabetic metabolic condition, or the discontinuation or modification of chemotherapy with a neurotoxic effect. Sharply severed nerves require surgical intervention. The patient's further fate depends on the capacity of the nerve for regeneration.

As already discussed, accompanying physical therapy is always indicated. The extent to which regeneration-promoting substances can be used is discussed below.

A number of medications have been used to promote regeneration, including thyroid hormones, cortisone, and testosterone. The severe side effects coupled with a lack of evidence of efficacy ended the use of these drugs. Alpha-lipoic acid, aldose reductase inhibitors for diabetes, and B vitamins are the only substances still discussed today, although their therapeutic use is by no means uncontroversial.

The role of B vitamins in the metabolism of the axon and their significance in myelin formation in Schwann cells are well known and are the reason why the vitamins are described as neurotrophic substances. It is undisputed in the discussion of the value of vitamin therapy that hypovitaminosis causes defined illnesses and that they can be cured by appropriate vitamin substitution. Classic hypovitaminosis from poor nutrition does not exist today. However, it is not rare to find B_{12} absorption disorders that require parental administration of vitamins. In INH treatment, a side effect of the drug is increased elimination of pyridoxine (B_6) and thus a vitamin deficiency polyneuropathy. Therefore, vitamin B_6 must always be substituted when INH treatment is undergone. It is also known that a vitamin B_1 deficiency can develop in alcoholics due to poor nutrition, the cause of Wernicke's encephalopathy. Prompt vitamin substitution can prevent irreversible brain damage in these cases. But administration of mega dosages of vitamins that are far beyond the body's needs and are immediately eliminated again is certainly useless, or as known in the case of vitamin B_6, can give rise to a toxic effect and cause a polyneuropathy.

It is not known exactly what dosages of what vitamins can be given when. The therapy situation for alcoholic polyneuropathy or other neuropathies that occur with or without a known underlying disease is thus unclear. Contributing to this is the fact that the determination of concentrations of vitamins in serum is insufficient for the diagnosis of a vitamin deficiency. In the end, the intracellular concentration of vitamins is significant. Vitamin deficiency, the classic precondition for therapy, is generally not discovered by testing vitamin levels. However, it is known that in physiological processes such as old age and pregnancy, a greater need for vitamins exists and normal vitamin levels lead to a relative deficit. It remains to be seen to what extent individual B vitamins become significant for regenerative processes in this situation and whether they can actually act as coenzymes to accelerate regenerative processes at higher concentrations. The lack of detailed information sustains the therapeutic expectations for vitamins and adds to their significance for future efforts.

10.3   Pain therapy

A pain state can be based on four mechanisms:
1. Peripheral nociceptive mechanism: a pathological process in the tissue that sustains a continuous stimulation or sensitization of nociceptors. Such a situation usually exists with musculoskeletal processes such as lumbago or other acute myofascial pain associated with inflammatory reactions.
2. Peripheral-central neuropathic mechanism: a peripheral nerve injury leads to pathological afferents in the dorsal horn. Such changes are found, for example, in polyneuropathy, deafferentation pain, herpes zoster, and trigeminal neuralgia.
3. Central neuropathic mechanism: pain syndromes as a result of damage to the central nervous system, such as thalamus pain after an ischemic insult or cerebral hemorrhage.
4. Psychogenic mechanisms: conditioning mechanisms such as morbid gain, psychosocial situation, stress-induced and others play a role.

Medical treatment must take into account that the patient's expectations are a key factor that codetermines treatment efficacy. An optimal therapeutic relationship is another important factor in providing a good basis for a positive therapeutic intervention. Careful assessment of a patient's expectations and previous therapy experiences should be evaluated when taking the medical history. Considering these aspect can help to include a placebo factor into the therapy and to avoid nocebo effects. In addition to these basic approaches, it is helpful to adhere to the following recommendations:

- The best medication for a specific patient has to be tested in the individual situation with respect to the underlying pain problem and possible pain mechanisms and with regard to drug-induced side effects and contraindications.
- The dose of the medication has to be titrated individually according to the therapeutic effect and side effects.
- Ineffective therapy can be judged only after 2–4 weeks of therapy under sufficient dosage.
- The application protocol and drug dosage have to respect the pharmacological profile of the drug and its possible mode of interaction.
- Fixed dose combination of analgesics with barbiturates, benzodiazepines, and caffeine or muscle relaxants should not be used as there is a constant risk of drug abuse or dependency.

Pain therapy can be necessary for an acute occurrence or be helpful in treating long-term, thus chronic, pain. Acute pain requires the systematic, fast, and immediately sufficient high-dosage administration of analgesics; sedation is usually not desirable. The procedure is different for chronic pain. Analgesics should be given only later if necessary and with reservations and only at low dosages. Treatment should be mainly with antidepressants or neuroleptics with sedating components. For chronic pain states, the use of non-pharmacological therapy forms is very important.

It is necessary to discuss the therapy options with the patient and to explain what can be achieved by the specific therapy. Nociceptive pain treatment is different from neuropathic treatment where the main focus of this chapter lies. Treatment of neuropathic pain is still difficult despite new treatments, and there is no single treatment that works for all conditions and their underlying mechanisms. With drug therapy, it is possible to reach tolerable pain reduction in about 50–80% of neuropathic pain states. Total pain relief is very rare. In 20–40% of cases, pain can be reduced only to a degree less than 50%, which may also be due to intolerable side effects. Realistic goals for the treatment of neuropathic pain are given below:
- Reduction of pain by > 50%
- Improvement of quality of life

- Improvement of sleep
- Maintain ability to work
- Maintain social activities and social relationships

In pain therapy, the preventive aspect is gaining more significance. It is an established fact that already existing, pronounced pain is difficult to treat and that it requires high doses of medication. For this reason, anticipatory pain therapy is propagated today. Prophylactic administration of low dosages of analgesics at an early stage creates freedom from pain that can be maintained in the further course of the illness with low dosages.

The efficacy of every drug must be distinguished from the ubiquitous effect of spontaneous healing or that of a placebo. Only drugs whose efficacy has been shown in a prospective, randomized, double-blind study should be viewed as therapeutically effective. If testing has not yet been completed for a drug, the therapist must assess it on his own. It must always be taken into consideration that in pain syndromes, drug therapy is only one aspect of treatment in addition to methods such as physical therapy, electrostimulation, and surgical, psychiatric, or psychosomatic intervention. Chronic pain usually requires a multimodal approach regarding diagnostics and treatment.

The required dosage, the dosage interval, and the form of application also require close supervision and must be oriented toward the pharmacological properties of the drug. Depending on the half-life of a medication, various saturation and maintenance levels must be used in treatment. If this aspect is not observed and the dosage is too low at the beginning, a sufficiently high therapeutic blood level cannot be reached quickly enough. This applies to antiphlogistics as well as non-steroidal anti-inflammatory drugs, usually abbrevated to NSAIDs. But gradual increase is often necessary for high dosages due to the side effects, especially with tricyclic antidepressants or anticonvulsives. When the therapeutic level has been reached, the effect need not set in immediately. For example, antidepressive substances do not become effective until a sufficient blood level has been reached. Then another 1 to 2 weeks are needed until it develops the clinically desired analgesic effect.

The kind of therapy planning is usually much more important than the selection of analgesic. Therapy planning also means that the treatment must be sufficiently long term and that there is good compliance and a cognizant patient.

10.3.1 Neuropathic pain treatment

There are no pathognomonic signs for neuropathic pain. Moreover, there is a great individual variability to developing chronic neuropathic pain. This seems to depend on genetic and environmental factors and their ease and likelihood to start and sustain peripheral or central sensitization. Before treatment, a careful clinical examination must be undertaken. The goal is to differentiate between nociceptive or neuropathic pain and the proportion of each type of pain when they co-occur . Electrophysiology and other techniques can help to further clarify the picture together with a psychological assessment and quantification of the amount of disability and disease-related deficits affecting the quality of life. In addition, questionnaires have been published to help with screening tools to distinguish between neuropathic and non-neuropathic pain. Their main strength lies in the fact that they can be used by non-specialists for screening and following up patients and assessing the efficacy of therapy. A test is only useful if it has been validated in the language it is applied. Some published tests are: pain DETECT, Short form of McGill Pain Questionnaire, Neuropathic Pain Questionnaire, and Neuropathic Pain Symptom Inventory or Pain Quality Assessment Scale. The severity of the patient's symptoms and pain can also be tested with the Michigan Neuropathy Screening Instrument, the Neuropathy Disability Score, or the Leeds Assessment of Neuropathic Symptoms and Signs. Furthermore, useful diaries have been developed in which patients note their pain type and intensity, the effects of pain on activities of daily living, and their fluctuations over time. The Zoster Brief Pain Inventory is a validated and convenient tool for the follow-up of patients with postherpetic neuralgia.

Recently, a systematic review and meta-analysis for the pharmacotherapy for neuropathic pain in adults was undertaken by the Special Interest Group on Neuropathic Pain of the IASP (Finnerup et al., *Lancet Neurol.* 2015; 162–173). Current knowledge and lines of evidence on pharmacotherapy have been published and recommendations are given for treatment and dose regimens (see Table 18). The proposed treatments do not consider the mechanisms underlying the neuropathic pain. Therefore, only general outlines are given. A strong recommendation for first-line medication is given as follows:

Gabapentin	1,200–3,600 mg, in three divided doses
Gabapentin extended release or enacarbil	1,200–3,600 mg, in two divided doses
Pregabalin	300–600 mg, in two divided doses
Serotonin-noradrenaline reuptake inhibitors Duloxetine venlafaxine extended release	60–120 mg, once a day 150–225 mg, once a day
Tricyclic antidepressants	Tapering 25–150 mg, once a day or in two divided doses

Low-level recommendations are given for the following medications:

Capsaicin 8% patches	One to four patches to the painful area of intact skin for 30–60 min every 3 months in peripheral neuropathic pain
Lidocaine patches	One to three patches to the region of peripheral neuropathic pain once a day for up to 12 h
Tramadol	200–400 mg, in two (tramadol extended release) or three divided doses
Botulinum toxin A	(Subcutaneously) 50–200 units to the peripheral neuropathic painful area every 3 months
Strong opioids	Individual titration as third-line possibility

The goal of modern pharmacotherapy is to tailor medication to the individual patient. For patients with neuropathic pain this means knowing the exact pathomechanisms maintaining the pain, irrespective of the etiology of the trauma. Patients with chronic neuropathic pain are notoriously difficult to treat. Knowledge of the underlying mechanism can help in the choice of an analgesic with an action to interfere with this mechanism. Such an individualized treatment is the aim of the "mechanism-based therapy." Unfortunately, the mechanism for some of the most frequent signs of neuropathic pain is not exactly known. Polypharmacy should be avoided. First one drug should be tested up to its pharmacologically tolerable limits. Patients respond differently to a given dose: some need high doses, others can be helped with low doses. If this regimen is not helpful, a second drug can be added.

Burning pain is regarded to be due to spontaneous activity in nociceptive-fiber pathways. Hyperexcitability induces this pain because of irritable nociceptors and regenerating nerve sprouts. Another mechanism can be denervation supersensitivity. In root avulsion, the second-order neuron becomes denervated and gives rise to spontaneous firing.

Demyelinated, non-nociceptive Aβ fibers seem to generate ectopic high-frequency bursts that cause shock-like sensations. Demyelination increases the susceptibility to ectopic excitation and high-frequency discharges. Demyelination exposes the neural membrane, which leads to exposed voltage-gated sodium channels. Another possibility is crosstalk to adjacent C fibers. These two mechanisms are believed to be causative in trigeminal neuralgia. Carbamazepine and oxycarbamazepine produce a frequency-dependent voltage-gated sodium channel block. This can block the ectopic and high-frequency discharges and is an adequate treatment for this disease.

The mechanisms underlying dynamic mechanical allodynia are still under investigation. Non-nociceptive stimuli might activate spared and sensitized nociceptive afferents. This painful situation is present clinically when in thoracic postherpetic neuralgia or sunburn the shirt touches the skin or when in diabetic polyneuropathy the feet come into contact with the bed sheets during the night.

In chronic pain it is known that psychological factors contribute to suffering and adaption in particular when treatment has not been successful (Re-

Table 18 Drugs for use in neuropathic pain therapy: summary of the results of a recent Cochrane review[a]

	First-line drugs			Second-line drugs			Third-line drugs	
	Serotonin-noradrenaline reuptake inhibitors duloxetine and venlafaxine	Tricyclic antidepressants	Pregabalin, Gabapentin, gabapentine extended relase	Tramadol	Capsaicin 8% patches	Lidocaine patches	Strong opioids	Botulinum toxin A
Balace between desirable and undesirable side-effects								
Effect size	Moderate	Moderate	Moderate	Moderate	Low	Unknown	Moderate	Moderate
Tolerability, Safety	Moderate	Low-Moderate	Moderate-high	Low-Moderate	Moderate-high	High	Low-Moderate	High
Values and preference	Low-Moderate	Low-Moderate	Low-Moderate	Low-Moderate	High	High	Low-Moderate	High
Strength of recommendation	Strong	Strong	Strong	Weak	Weak	Weak	Weak	Weak
Neuropathic pain condition	All	All	All	All	Peripheral	Peripheral	All	Peripheral
Quality of evidence	High	Moderate	High	Moderate	High	Low	Moderate	Moderate

a The recommendations help in selecting the medication as monotherapy or, if indicated, in combination.

view: Vase et al Pain 2016; 157 S98-S105). The way of prescribing and dispensing medication contributes to the therapeutic effect. The perception of receiving a treatment does not only contribute to the efficacy of inert treatment but also to the efficacy of active treatment. The placebo effect underlines the importance of psychological factors. As pain levels fluctuate over time, it is important to differentiate between the natural history of the pain, the placebo response, and the placebo effect. The patient's expectations of a treatment effect are not neutral but embedded in emotional feelings. The placebo response is derived from the patient's positive perception and experience of receiving a pain-reducing treatment and as a result a pain-reducing effect. A nocebo effect will arise after administration of an inert treatment along with negative behavioral procedures or verbal suggestions that tend to worsen the pain. The doctor-patient interaction and how a therapeutic regimen is explained or given to a patient thus play an important role.

Social factors can also influence the amount of pain. Patients who display few objective signs of a disease but still report intense pain are sometimes involved in injury benefit claims. The expected compensation can make a treatment that requires compliant participation difficult.

Chronic pain is not easy to treat and reported pain intensity should not always be an indicator of the amount of pain-killing drugs to be taken. Chronic pain is not primarily determined by nociception. The impact of pain on the quality of life is more strongly associated with treatment preference than is pain intensity. Therefore, it is important to evaluate social, psychological, and psychiatric and other co-morbid parameters. Treatment recommendation should be based on the knowledge of the patient's function, satisfaction with care, and quality of life.

10.3.2 Non-steroidal anti-inflammatory drugs

Inflammation is an extremely common reason for people to seek medical treatment. Current therapies for inflammatory diseases, primarily consisting of steroidal and NSAIDs, immunosuppressive agents, and recently, biological therapies (e.g., anti-TNF-α)

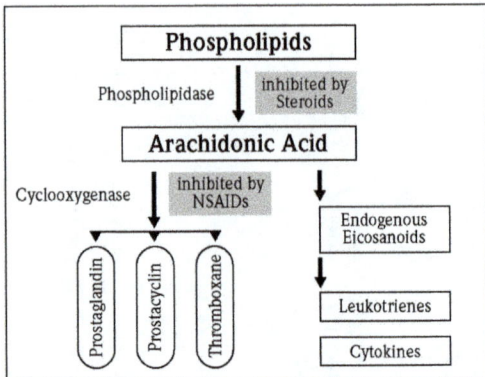

Figure 102 Scheme of biochemical substances which play a role in inflammation and pain. The breakdown of the phospholipid membrane by enzymes constitutes a cascade of events. In the end, these processes trigger substances that elicit pain (Figure 89). Pharmacological inhibition of the enzymes can stop this cascade and pain

provide only temporary relief, only partially ameliorate the disease process, and have significant adverse effects. An appropriate noxious stimulus liberates arachidonic acid from the storage lipid and converts it to prostaglandin, a powerful pain-producing substance. The arachidonic acid in these lipids is released in an enzymatic reaction catalyzed by phospholipase A$_2$ (Figure 102).

NSAIDs are drugs with analgesic, antipyretic, and anti-inflammatory effects. The name "non-steroidal" is used to distinguish these pharmacological substances from steroids, which have a similar eicosanoid-depressing, anti-inflammatory action. As analgesics, NSAIDs are unusual in that they are non-narcotic.

Eicosanoids, derived from 3-omega fatty acids, are signaling molecules that exert control in inflammation and immunity. Anti-inflammatory drugs such as aspirin and other NSAIDs act by down-regulating eicosanoid synthesis. There are four families of eicosanoids – the prostaglandins, prostacyclins, the thromboxanes and the leukotrienes. Cyclooxygenase (COX) is an enzyme that is responsible for the formation of important biological mediators called prostanoids which include the prostaglandins, prostacyclins, and thromboxanes. Pharmacological inhibition of COX with NSAIDs can provide relief from the symptoms of inflammation and pain; this is the method of action of NSAIDs, such as the well-known aspirin and ibuprofen. These are powerful drugs that can slow down or stop neurogenic inflammation (see Sect. 6.4).

Classification of NSAIDs can be based on their chemical structure and within a group they tend to have similar characteristics and tolerability. In a comparison of individual drugs there is little difference in their clinical efficacy when used at equivalent doses. The class of NSAIDs is composed of salicylates (acetylsalicylic acid), arylalkanoic acids (diclofenac, indomethacin), profens (ibuprofen), pyrazolidine derivatives (phenylbutazone, metamizole), oxicams (piroxicam), and the COX-2 inhibitors (celecoxib). Paracetamol (acetaminophen) has negligible anti-inflammatory activity, and is strictly speaking not an NSAID. There is also some speculation that paracetamol acts through the inhibition of the recently discovered COX-3 isoform.

Due to the pain-releasing effect of NSAIDs, COX is of great scientific interest. Two isoforms exist, COX-1 and COX-2. COX-1 is an enzyme with a "housekeeping" role in regulating many normal physiological processes such as blood flow in the tissue. COX-2 is not normally present in the cell but is up-regulated during inflammatory situations by various inflammatory mediators such as cytokines. It is an inducible enzyme, becoming abundant in activated macrophages and other cells at sites of inflammation. For the site of action it is speculated that NSAIDs may act centrally, possibly within the spinal cord. However, the mechanism of action in this case is not well characterized.

Drugs that selectively inhibit COX-2 activity (the coxibs) were therefore believed to have better therapeutic value than the non-selective NSAIDs. Although the anti-inflammatory and analgesic capacity of coxibs is convincing, it has been shown that some of them produce severely hazardous side effects such as myocardial infarction, hypertension, and chronic renal failure. If a COX-2 inhibitor is taken, one should not use a traditional NSAID concomitantly. In addition, patients on daily aspirin therapy (as for reducing cardiovascular risk or colon cancer risk) need to be careful if they also use other NSAIDs, as the latter may block the cardioprotective effects of aspirin. It was hoped that the COX-2

would reduce adverse gastrointestinal drug reaction, but studies have shown no significant difference when compared with diclofenac.

Acute inflammation can be initiated by a number of inflammatory triggers. This will further result in a programmed sequence of physiological mechanisms that begin with the release of TNF-α, IL-1β, and IL-6. Like many other substances, cytokines are proinflammatory in agonistic interaction with phospholipases and prostaglandins. These cytokines activate a cascade of responses that result in the release of cortisol, which dampens inflammation by down-regulating the release of TNF-α, IL-1β, and IL-6. So corticosteroids are the main effector endpoint of the neuroendocrine immune response to inflammation. As can be seen in Figure 99, steroids are capable of suppressing the cascade of reactions in inflammation. This is the rationale for the therapeutic application of glucocorticoids, as they have an indirect inhibitory effect on the activity of phospholipase, the enzyme that plays an essential role in the synthesis of prostaglandins and leukotrienes. Leukotrienes are produced in the body from arachidonic acid by the enzyme 5-lipoxygenase. Leukotrienes are very important agents in the inflammatory response. By their chemotactic effect they help to bring the necessary cells to the tissue. Leukotrienes also have a powerful effect in vasoconstriction and they can increase vascular permeability.

Cytokines also induce enzymatic reactions catalyzed by phospholipase A_2, which in turn activates the inflammatory cascade. In the future, substances that regulate cytokine release in the peripheral and central nervous system may become treatment options for influencing inflammatory-neuropathic pain. In fact, interesting insights are already available today. Studies in patients with depression showed a lowering of TNF-α serum levels in amitriptyline recipients. An immunomodulating effect has also been postulated for anticonvulsant drugs and opioids. The focus is on cytokines, since research has shown that they can play a role in painful root lesions. Until recently it was believed that intervertebral discs play only a limited role in immunological processes, as a result of which the production and secretion of cytokines in intervertebral discs would be possible to a limited extent only, if at all, due to a lack of immunocompetent cells in intervertebral disc structures. In animal studies it was shown that cytokines are involved as mediators of an inflammatory-neuropathic component in sciatica: the epidural administration of autologous nucleus pulposus material causes intraneural edema and reduction of nerve cell conduction velocity in rats. Also, the expression of TNF-α was demonstrated in herniated intervertebral discs and its application to nerve roots caused neuropathic alterations.

In cases of a mixed pain syndrome, NSAIDs have only a limited effect – they can influence the nociceptive pain component but are ineffective for the treatment of the neuropathic pain.

10.3.3 Anticonvulsants

Anticonvulsants are a treatment option in neuropathic pain states when the underlying pathophysiology involves membrane instability and displays clinical features of ectopic impulse generation and ephaptic spread of impulses. As described in Sect. 7.3, membrane instability results from myelin damage and the expression of new receptors and ion channels on the axons and neuron membrane. Pregabalin binds to the subunit of a tension-dependent calcium channel and as such may have an impact on pathological impulse propagation and central sensitization processes. The effect of both substances is mediated only to a minor extent by interaction with GABA receptors. Pregabalin has a modulating action on a subunit of Ca channels, especially on those located presynaptically at the nociceptive terminal in the dorsal horn of the spinal cord. This reduces the release of glutamate and substance P.

In trigeminal neuralgia and post-stroke pain, lamotrigine in doses up to 400 mg daily has a pain-relieving effect. A positive therapeutic effect has also been reported for carbamazepine and phenytoin. Carbamazepine and oxcarbazepine are sodium channel blockers. Oxcarbazepine is reported to have fewer side effects than carbamazepine, but with comparable analgesic effects. Judging by the clinical impression, they are particularly effective in controlling shooting or triggered pain. Valproate in dosages up to 1,200 mg proved to be helpful in painful diabetic neuropathy and postherpetic neuralgia.

In all cases, the dosages of anticonvulsants must be titrated during therapy from small initial doses in incremental steps to higher doses. The speed of titration depends on the side effects, the severity, and acuity of the pain problem and whether a patient is hospitalized or on an outpatient treatment regimen. The only valid parameter of the effectiveness of the therapy is the therapeutic result; side effects preclude further dosage increments. Monitoring plasma concentrations is not very helpful for the acute therapy but can indicate the patient's compliance in longer-term treatments.

10.3.4 Antidepressants

Tricyclic antidepressants have analgesic properties of their own; pain reduction seems therefore not to be based on an antidepressant effect. Antidepressants induce their analgesic effect via their action on descending modulatory inhibitory controls. We know today that this group of substances reduces the reabsorption of monoamines, such as serotonin, in cells of the central nervous system and thus inhibits synaptic functions. These substances also block tension-dependent potassium channels and have sympatholytic properties. In addition, antidepressants can bind to central receptors that play a role in the complex system of pain modulation. This is the mechanism behind the effect of these biogenic amines that are present in the pain-modulating descending pathway systems. Amitriptyline, desipramine, and nortriptyline have proven to be effective in the therapy of chronic neuropathic pain, especially for diabetic neuropathies, myofascial syndromes, postherpetic neuralgia, and atypical facial and head pain. The dosages must begin gradually – as a rough indication, initial low dosages of about 20 mg at night are recommended, which can be increased after a few days to 50–75 mg up to around 50–150 mg. The mean dose required for pain reduction is below the antidepressant dose and pain reduction occurs within days to 2 weeks, while antidepressant effects take several weeks to develop at higher doses.

The manufacturers' information should be observed when dealing with the individual drugs.

The serotonin and norepinephrine reuptake inhibitor amitriptyline is currently the best studied substance. Several controlled studies have demonstrated a positive effect in postherpetic neuralgia and diabetic neuropathy. All neuropathic pain types such as burning spontaneous pain, attacks of shooting pain, and allodynia can be suppressed.

Newer antidepressants are neither selective serotonin reuptake inhibitors (SSRI) nor tricyclic antidepressants. Venlafaxine (a mixed serotonin noradrenaline reuptake inhibitor) and duloxetine both block serotonin and norepinephrine reuptake and were also effective in treating neuropathic pain in painful diabetic neuropathy.

10.3.5 Opioids

Opioids are discussed here only in the context of neuropathic pain. In nociceptive pain, they play an important role for pain relief. But their role in the context of neuropathic pain is still under discussion. Opioids interact with the μ-opioid receptor in the central nervous system. A weak opioid is tramadol with serotonin and norepinephrine reuptake inhibition. Tapentadol only has norepinephrine reuptake inhibition. The drugs have shown moderate effect in nociceptive pain but not in musculoskeletal pain like low-back pain. Intravenous opioid administrations have been given in peripheral neuropathic pain and for mixed pain conditions and showed a positive effect. So there are positive indicators that opioids are effective also in neuropathic pain. Yet opioids have many side effects and there is the possibility of developing tolerance and physical dependency. All of these factors limit use of the drugs in daily routine. Morphine has been effective in phantom limb pain. Tramadol proved to be effective in painful polyneuropathy and postherpetic neuralgia, but in this study there was also a very high placebo responder rate, which shows that well-designed further studies are necessary to give proof for the appropriate indications for these drugs. It seems that opioids are frequently given in mixed pain states such as back pain or unclear states such as fibromyalgia, where there is actually no indication and the whole array of other possible interventions was not appropriately implemented. Aside from the analgesic effects, opioids have hyperalgesic properties when given in high doses. So it is not rare to see

patients suffering from musculoskeletal pain who do not profit from opioids even under increasing doses. In this case it is worthwhile to withdraw opioids in order to see if they are indicated at all and if they are possibly overdoses. For these cases it is also established experience that if a test trial with opioids at a sufficient dose does not show an effect, they should not be given as they have not proved to be helpful for this specific case and pain state.

Opioids in non-malignant pain should be given on a long-term basis only when there has been interdisciplinary pain counseling. The appropriate dose can then be found only by incremental titration. The therapy must be guided by a physician who has experience with opioid therapy.

10.3.6 Neuroleptics

In pain therapy, the sedating effect of neuroleptics is often exploited, but an analgesic effect has not been proven for this substance group. Sedation with neuroleptics appears to be helpful, since tranquilizers, which are used for chronic pain over long therapy periods, have a high potential for addiction and dependency. The mode of action of neuroleptics for pain, especially tumor pain, has not yet been explained. Opiate antagonist effects appear to play a role by binding to central opiate receptors. In pain therapy, haloperidol can be used (1 ml = 20 drops contains 2 mg); initially doses of 0.5–1 mg daily are given.

10.3.7 Miscellaneous

Substances such as baclofen are not useful for treating local muscle spasms, but to reduce a centrally elevated spastic muscle tone. Antispasmodics such as benzodiazepine, tetrazepam, tizanidine, flupirtinemaleate and others are praised as adjuvant therapy, especially for myofascial syndromes. While the substances show tonolytic activity in animal tests, it is uncertain to what extent they can actually have the desired therapeutic effect in humans. In addition to pronounced side effects on the central nervous system, the tolerance effect and a potential for developing dependency must be kept in mind for some of the substances and for long-term treatment. As noted in Sect. 5.1.5, baclofen can be used to treat trigeminal neuralgia.

In postherpetic neuralgia with allodynia, topical lidocaine could reduce the pain. As add-on therapy the lidocaine patch can also be used in peripheral neuropathies with mechanical allodynia. The local anesthetic induces an unspecific blockade of sodium channels in the membrane of peripheral nerves.

Oromucosal cannabinoids have been studied for central pain states such as post stroke and multiple sclerosis. Tetrahydrocannabinol dronabinol in a dosage of 5–10 mg/day has been reported to relieve pain in multiple sclerosis patients. For neuropathic pain, this drug showed no positive effect. Their value is still under controversial debate.

Local application of capsaicin is helpful in painful diabetic neuropathy and postherpetic neuralgia. It is an ingredient of the hot red pepper. Its function in nociception is to bind to a specific receptor, called the vanilloid receptor (TRPV1). The first cloned nociceptive ion channel was the TRPV1 receptor, which is expressed in about 40% of DRG cells. This ion channel is opened by binding of capsaicin, the compound in hot pepper that causes burning pain. These channels are predominately expressed in small sensory C fibers and to a lesser extent in Aδ fibers, both of which terminate in the spinal dorsal horn, where TRPV1 is localized to both pre- and postsynaptic neurons as well as to glial cells. TRPV1 is a critical integrator of inflammatory pain signaling. TRPV1 is activated by heat (>43 °C) as well as by endogenous eicosanoids and protons. The TRPV1 channel is thought to be a primary mediator of warm thermal sensation. The receptor functions as a ligand-gated channel, allowing an influx of a large amount of calcium and a small amount of sodium. In addition to activating pain pathways, the vanilloid receptor also triggers the release of peptides, leading to an inflammatory response. Capsaicin is an alkaloid that depletes tissues of substance P (SP) and reduces neurogenic plasma extravasation, the flare response, and chemically induced pain. SP is present in afferent neurons innervating skin, mainly in polymodal nociceptors, and is considered the primary neurotransmitter of painful stimuli from the periphery to the central nervous system.

For polyneuropathy, capsaicin must be applied onto the painful skin area four times a day for 4–6 weeks. At first there is a painful burning sensation due to afferent C fiber stimulation and histamine release from mast cells. This sensation can be reduced by first using a lidocaine patch. Over time, the burning sensation ceases due to SP depletion.

Vitamins B_1, B_6, and B_{12} have been shown to have a pain-relieving effect and as an add-on therapy can enhance the efficacy of gabapentin or diclofenac in mixed pain states.

Cyanocobalamin (B_{12}) shows a dose-dependant anti-allodynic effect in neuropathic pain, whereas diclofenac did not show relevant efficacy at doses effective in models of nociceptive pain. Anti-allodynic action was also seen in B_1 application alone; however, anti-allodynic action is also detectable when using a mixture of all three vitamins. The B vitamins enhance and extend the efficacy of drugs such as gabapentin and diclofenac.

The main indication for neurotoxin botulinum toxin type A is the treatment of focal muscle and tone hyperactivity. The toxin was used In some peripheral neuropathies and chronic headache, and an analgesic effect was reported. It may be that the toxin interferes with inflammation and exerts an analgesic effect.

10.3.8 The mixed pain concept and its impact on pain therapy

Evaluating, differentiating, and characterizing the underlying mechanisms of chronic pain syndromes are crucial, not only for scientific relevance, but for therapeutic reasons. Neuropathic pain may require an entirely different pain management strategy than nociceptive pain. In recognition of the variety of mechanisms involved in the different pain components, the term mixed pain syndrome has come to be used to describe a combination of nociceptive, inflammatory, and neuropathic pain components. Therefore, the mixed pain concept appears to be particularly apt for describing musculoskeletal problems such as back pain and sciatica. In neuropathic pain, the lesion must involve the somatosensory system with its ascending and descending pathways. It is the clinician's task to clarify whether physiological pain is involved, which induces changes in the nociceptive system due to inflammation or musculoskeletal pain.

These are pain states with a high prevalence in the population, a high socio-economic impact, and a great need for appropriate pain therapy. The pathophysiological concepts clearly indicate that back pain is made up of a number of pain components with differing pathophysiological mechanisms. Localized nociceptive back pain can be differentiated from several neuropathic pain components mechanisms and the nociceptive pain can also trigger and sustain neuropathic pain.

A study among doctors treating patients with low back pain showed that the overwhelming majority of medication was directed against the nociceptive pain component and neuropathic pain was only rarely considered. First-line medications were usually NSAIDs and in severe cases opioids, which work on the basis of an anti-inflammatory and analgesic mechanism. But it has to be stressed again that NSAIDs have no effect on the neuropathic pain components. This may be one of the reasons why patients with low back pain develop chronic pain states and are disabled in their private life and on the job. Co-analgesics such as anticonvulsants and antidepressants are indicated and should be prescribed for these pain components. These agents act on channel proteins and receptors expressed only after a nerve injury. This explains their specific effect in neuropathic pain and their lack of effect where nerve fibers are intact. B vitamins can extend the efficacy of NSAIDs and coanalgesics. Drug treatment with drugs of proven efficacy should be considered in patients with sciatica and low back pain. This applies in particular to patients who have not responded to prior treatment attempts and who are in danger of developing a chronic pain condition.

Subject Index

CPI Antony Rowe
Chippenham, UK
2018-03-28 21:27